CAUSAL INFERENCE

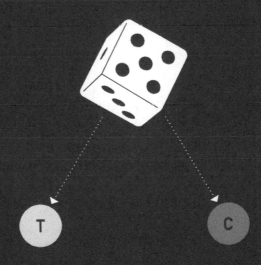

The MIT Press Essential Knowledge Series

A complete list of books in this series can be found online at
https://mitpress.mit.edu/books/series/mit-press-essential-knowledge-series.

CAUSAL INFERENCE

PAUL R. ROSENBAUM

The MIT Press | Cambridge, Massachusetts | London, England

The MIT Press would like to thank the anonymous peer reviewers who provided comments on drafts of this book. The generous work of academic experts is essential for establishing the authority and quality of our publications. We acknowledge with gratitude the contributions of these otherwise uncredited readers.

This book was set in Chaparral Pro by New Best-set Typesetters Ltd. Printed and bound in the United States of America.

Library of Congress Cataloging-in-Publication Data

Names: Rosenbaum, Paul R., author.
Title: Causal inference / Paul R. Rosenbaum.
Description: Cambridge, Massachusetts : The MIT Press, [2023] | Series: The MIT Press essential knowledge series | Includes bibliographical references and index.
Identifiers: LCCN 2022022398 (print) | LCCN 2022022399 (ebook) | ISBN 9780262545198 (paperback) | ISBN 9780262373531 (epub) | ISBN 9780262373548 (pdf)
Subjects: LCSH: Causation. | Science—Experiments. | Observation (Scientific method) | Inference. | Probabilities. | Mathematical statistics.
Classification: LCC Q175.32.C38 R663 2023 (print) | LCC Q175.32.C38 (ebook) | DDC 001.4/2—dc23/eng20221004
LC record available at https://lccn.loc.gov/2022022398
LC ebook record available at https://lccn.loc.gov/2022022399

10 9 8 7 6 5 4 3 2

In learning the nature of things, we learn the possibilities of action; in learning the possibilities of action, we learn the nature of things.

—Peter F. Strawson, *Analysis and Metaphysics*

To find out what happens to a system when you interfere with it, you have to interfere with it.

—George E. P. Box, "Use and Abuse of Regression"

CONTENTS

SERIES FOREWORD

The MIT Press Essential Knowledge series offers accessible, concise, beautifully produced pocket-size books on topics of current interest. Written by leading thinkers, the books in this series deliver expert overviews of subjects that range from the cultural and the historical to the scientific and the technical.

In today's era of instant information gratification, we have ready access to opinions, rationalizations, and superficial descriptions. Much harder to come by is the foundational knowledge that informs a principled understanding of the world. Essential Knowledge books fill that need. Synthesizing specialized subject matter for nonspecialists and engaging critical topics through fundamentals, each of these compact volumes offers readers a point of access to complex ideas.

LIST OF EXAMPLES

- Bleeding George Washington (chapter 1)

- Treatments for Ebola (chapter 2)

- Smoking and periodontal disease (chapters 3–5)

- Smoking and lung cancer (chapters 3, 5, and 8)

- Lethal side effects of antibiotics (chapter 6)

- The Earned Income Tax Credit and employment (chapter 6)

- Winning the lottery and going bankrupt (chapter 7)

- Housing subsidies and employment (chapter 7)

- Genetic causes of Graves' disease, Alzheimer's disease, and autism (chapter 7)

- Do unions raise wages? (chapter 7)

- Mindfulness and smoking (chapter 7)

- Clinical treatment of addiction to heroin and cocaine (chapter 8)

- Bringing workplace toxins home to your children (chapter 8)

- Does a glass of red wine lengthen or shorten your life? (chapter 9)

LIST OF METHODOLOGICAL TOPICS

- Definition of causal effects (chapter 1)

- Randomized experiments (chapter 2)

- Boxplots explained (chapter 3)

- Propensity score (chapters 3–4)

- Sensitivity analysis (chapter 5)

- Multiple control groups (chapter 6)

- Counterparts (chapter 6)

- Sibling tests (sib-TDT) for genetic causes of disease (chapter 7)

- Parent-child tests (TDT) for genetic causes of disease (chapter 7)

- Instrumental variables (chapter 7)

- Evidence factors (chapter 8)

- Mendelian randomization (chapter 9)

THE EFFECTS CAUSED
BY TREATMENTS

In 1757, before he defeated an empire, before he rose to continental power and peacefully transferred that power following an election—before all of that—George Washington was ill, and his doctors nearly killed him. They bled him to restore a healthy balance in his humors. Much later, on December 13, 1799, Washington

> complained of a sore throat . . . so sore he could scarcely breathe. . . . Not only was the dying man slowly choked to death by the swelling in his throat, but he had to endure the tortures of eighteenth-century medicine. He was repeatedly bled until he had lost as much as half his blood volume. He was forced to vomit copiously. . . . Burning chemicals were placed on his skin to blister him.[1]

Washington died on December 14, 1799.

What are humors? Understood by the Greek physicians Hippocrates and Galen, imbalances in humors caused disease, and the physician's job was to assist the patient in restoring the balance. In classical Greek thought, there were four elements—earth, water, air, and fire—and these were represented in the body as four humors—black bile, phlegm, blood, and yellow bile.[2] The imbalances in humors were evident—always present in constant conjunction—in disease. The ill were too hot or too cold—had fevers or chills; they coughed, sneezed, and gasped for air as they tried to rebalance their gaseous humors; they released blood, pus, phlegm, vomit, and diarrhea; they were swollen or emaciated. No one returned to health until these imbalances in humors were corrected.

The theory of humors mistook the symptoms for the disease, the conspicuous for the consequential, the collateral for the cause. Doctors believed in humors because their teachers believed in humors, as did those who had taught their teachers. Imbalances in humors caused disease—that was believed for more than twenty centuries. Two thousand years.

Why did eighteenth-century physicians bleed their patients? A plausible but mistaken answer is that they knew little biology. True, they knew little biology. The discovery of DNA would occur in the twentieth century. Charles Darwin theorized about evolution by natural selection in the nineteenth century. The importance of microorganisms to disease would be discovered in the nineteenth century by Louis Pasteur, Jacob Henle, and Robert Koch. Pasteur began his biological work in an open-minded but utterly serious industry with no tolerance for faulty reasoning, inaccurate measurement, or mistaken theories; he was a French chemist engaged in the production of fine wine.[3] Better theory meant better wine; you could taste it.

True, the eighteenth century knew little biology, but you do not need to know biology to know that bleeding patients is harmful. You can know with complete rigor the effects caused by a treatment with no understanding of the biological mechanism through which those effects are produced. Indeed, if your theories of disease promote treatments that kill your patients, perhaps that is a good moment to reexamine your theories. Eighteenth-century medicine lacked what John Dewey would call the "experimental habit of mind," together with certain methodological developments of the twentieth century, principally the development of randomized experiments.[4]

What Is a Causal Effect? Why Is Causal Inference Difficult?

Did bleeding Washington cause his death? Would bleeding patients often cause their deaths? An important methodological discovery of the twentieth century says that the second question has a rigorous answer, while the first is, and will remain, a matter of speculation. The rigorous answer is the topic of chapter 2, but the first task is to understand the two questions, how they are different, and why they might have different answers.

To ask whether bleeding Washington caused his death is to imagine and compare two worlds. In one world—our world, the actual world—Washington was bled in 1799 and died the following day. In the other world, Washington was not bled. Would Washington have survived in that world in which he was not bled? If Washington would have died anyway from complications of his sore throat, then bleeding did not cause his death. If he would have survived in that other world, then bleeding caused his death. The causal effect of bleeding on Washington's survival is a comparison of his survival in two worlds, one in which he was bled, and the other in which he was not. The basic issue is that we saw what happened in the actual world—he was bled and died—but we cannot see what would have happened to him in the alternative world in which he was not bled. How are we to learn about a world we cannot see?

This basic issue is one problem, but there are others. Washington's alternative world seems made up—a fiction, a matter of stipulation, not experimentation. What else happens in this alternative world? How does it differ from the actual one? Are all the elements of eighteenth-century medicine withheld, and his sore throat allowed to run its course? Or is bleeding the only treatment that is withheld? Is there anything real about this alternative world if it is up to you or me to stipulate what is or isn't true in that alternative world? In chapter 2, the alternative worlds are actual, specified in an experimental protocol, worlds from which data are extracted; the possible worlds are as real and definite as worlds can be.

Notation for a Precise Question

A rigorous solution is preceded by a precise statement of the problem. A precise statement is a precise statement; it does not solve a problem. A precise statement is a framework; within it, a problem is either solved, or shown to have no solution or to be incoherent. Washington's fate, had he not been bled, has no solution, but whether bleeding patients is harmful has a solution. To see this, we must ask these questions in a precise language.

Mathematician George Pólya wrote,

An important step in solving a problem is to choose the notation. It should be done carefully. The time we spend now on choosing notation may well be repaid by the time we save later by avoiding hesitation and confusion. . . . A good notation should be unambiguous, pregnant, easy to remember; it should avoid harmful second meanings, and take advantage of useful second meanings; the order and connection of signs should suggest the order and connection of things.[5]

The effect of bleeding on Washington's survival is a comparison of his fate in two worlds, one in which he is bled, and the other in which he is not. Let's call these worlds T and C, for treatment and control. In world T, Washington is bled, but in world C he is not bled. We are concerned with Washington's survival in these two worlds. Let us use 1 for survived to January 13, 1800, and 0 for died before January 13, 1800; so we are speaking about survival for one month beyond the start of the treatment. We know Washington died promptly having been bled. Would Washington have survived a month if he had not been bled?

We are speaking about Washington, w, and his fate in two worlds, T and C. The symbol r_{Tw} is Washington's survival in world T, in which he was bled: $r_{Tw} = 1$ if he survived to January 13, 1800, and $r_{Tw} = 0$ if he died before January 13, 1800. As it happened, we saw world T—Washington was bled—and we saw that he died, $r_{Tw} = 0$. In parallel, the symbol r_{Cw} is Washington's survival in world C, in which he was not bled: $r_{Cw} = 1$ if he would have survived to January 13, 1800, and $r_{Cw} = 0$ if he would have died before January 13, 1800. A causal effect is a comparison of outcomes in two possible worlds, a comparison of r_{Tw} and r_{Cw}, such as $r_{Tw}-r_{Cw}$. Here, $r_{Tw}-r_{Cw} = 0 - 1 = -1$ if Washington would have survived had he not been bled, $r_{Tw}-r_{Cw} = 1 - 0 = 1$ if he would have survived only if he had been bled, and $r_{Tw}-r_{Cw} = 0$ if his fate would have been the same regardless of the treatment.

The dilemma is clear: we saw $r_{Tw} = 0$—Washington died having been bled—but we did not see r_{Cw}—we did not see what would have happened had he not been bled—so we cannot determine whether $r_{Tw}-r_{Cw} = 0 - 1 = -1$ or $r_{Tw}-r_{Cw} = 0 - 0 = 0$. In other words, we cannot calculate the causal effect, $r_{Tw}-r_{Cw}$, because it involves something we did not see, namely r_{Cw}. This notation expresses the effects caused by treatments as comparisons of potential outcomes under alternative treatments, and it is due to statisticians Jerzy Neyman and Donald Rubin.

Would a Control Group Solve the Problem?

A first thought—not entirely wrong, yet not correct either—is that we just need a control group. What is a control group? Why is a control group, by itself, inadequate when estimating the effects caused by treatments? What else is needed if a control group is to be helpful? Consider the simplest case: two individuals, one treated, and the other a control. In place of one individual, Washington or

w, there are two individuals, Kim or k, and James or j. As was true of Washington, Kim has two potential outcomes, r_{Tk} and r_{Ck}, and so does James, r_{Tj} and r_{Cj}. The effect of the treatment on Kim compares her two potential outcomes, $r_{Tk}-r_{Ck}$, and because Kim receives one treatment or the other, we cannot see both r_{Tk} and r_{Ck}, so we cannot calculate her causal effect, $r_{Tk}-r_{Ck}$. All of this is true for James as well: he receives either treatment or control, so we see either r_{Tj} or r_{Cj} but not both, and thus we cannot calculate his causal effect, $r_{Tj}-r_{Cj}$.

At first, it seems we have two Washington problems, one for Kim and one for James, and two problems is two problems, not a solution. Suppose, however, that Kim receives the treatment and James receives the control. Does that help? Then we see Kim's response to the treatment, r_{Tk}, and James's response to the control, r_{Cj}. So on second thought, that does seem relevant, better than the Washington problem; at least we saw the outcome under each treatment. True, we have just Kim and James, and two people does not seem to be enough, but if causal inference just needed more people, then there is no shortage of them. We will need more people, but more people is not enough. Causal inference is not about bigger data; it is about better data.

Certainly, our second thought—that there is benefit to seeing people under both treatment T and treatment C—has dissolved one aspect of the Washington problem.

The possible world of the control C is now a fact, not a fiction. No stipulation is involved in the definition of the control condition C if people are actually receiving the control C.

Our plan has been to give some people treatment T and other people control C, and then compare the outcomes of people who received treatments T and C. If there were many people in each treatment group, then we might compare the average outcomes in each group. With two people, comparing the averages simplifies to comparing Kim's outcome to James's outcome. If Kim is given T and James is given C, then we estimate the effect of treatment as Kim's response under T, namely r_{Tk}, minus James's response under C, namely r_{Cj}, so the estimate is $r_{Tk}-r_{Cj}$, which has the distinct attraction of being the difference of two quantities that we observed. Much may be wrong with $r_{Tk}-r_{Cj}$, but it is arithmetic, not metaphysics. In parallel, if Kim is given C and James is given T, then the estimated effect is $r_{Tj}-r_{Ck}$, which is again the difference of two quantities that we have observed. Does this work?

Alas, no. To see why it does not work, suppose that there is no difference between treatments T and C for Kim, $r_{Tk}=r_{Ck}$ or $0=r_{Tk}-r_{Ck}$, so Kim's fate is the same whether she receives T or C. To be definite, suppose that Kim would survive under T and C, so $1=r_{Tk}=r_{Ck}$ and $0=1-1=r_{Tk}-r_{Ck}$. Suppose also that there is no difference between T and C for James, $r_{Tj}=r_{Cj}$ or $0=r_{Tj}-r_{Cj}$, but unlike Kim, James would die under both treatments. In brief, we are supposing that giving treatment T rather than C does nothing for either Kim or James, but Kim and James are themselves different, and face different fates regardless of what treatment they receive.

The difficulty is evident. Kim and James are different people, with different outcomes, so comparing Kim under the treatment to James under the control is not estimating anyone's causal effect, neither Kim's effect $r_{Tk}-r_{Ck}$ nor James's effect $r_{Tj}-r_{Cj}$. The treatment has no effect on Kim or James, so we are trying to estimate zero effect, because $0 = 1 - 1 = r_{Tk}-r_{Ck}$ for Kim and $0 = 0 - 0 = r_{Tj}-r_{Cj}$ for James. If we give treatment T to Kim and treatment C to James, the estimate of the effect is not zero, but rather $r_{Tk}-r_{Cj} = 1 - 0 = 1$, so treatment T misleadingly appears to be the perfect treatment: it appears to save everyone. Had we given C to Kim and T to James, then treatment T looks awful, $r_{Tj}-r_{Ck} = 0 - 1 = -1$: it appears to kill everyone.

In the imagined situation, the average treatment effect (ATE) on Kim and James is, of course, zero: it is the average of $0 = r_{Tk}-r_{Ck}$ for Kim and $0 = r_{Tj}-r_{Cj}$ for James, and the average of 0 and 0 is $(0 + 0)/2 = 0$. To calculate the average treatment effect, we would have to see Kim's fate under both treatment T and C, and also James's fate under both T and C, but the problem all along has been that we can see neither $r_{Tk}-r_{Ck}$ for Kim nor $r_{Tj}-r_{Cj}$ for James, so we cannot calculate the average treatment effect.

It Is Just the Same with More Than Two People

So far, we have talked about one person, Washington or w, or two people, Kim and James or k and j. Nothing much changes if there are more than two people. In chapter 2, there will be 343 people. Suppose that there are I people, where I is some number, such as $I = 343$, as in the next chapter. Naturally, we number the individuals from 1 to 343, or generally from 1 to I. We use a lowercase i to refer to anyone, in much the same way we use a pronoun. We do not want to have to say that something is true of person 1 and also person 2, and even person 3, and on and on, until finally we say it is true of person 343 too, because if we talked that way about 343 people then you would not be able to stand it, and neither would I. So instead we say simply that something is true of person i for $i = 1, 2, \ldots, I$. That is, we say it is true for person i no matter who person i happens to be.

A causal effect is a comparison of the outcome a particular person would exhibit if given the treatment and the outcome this same individual would exhibit if given the control condition. Causal inference is challenging because, for any individual, we never see both outcomes. For any individual i, the response of i if treated is r_{Ti}, and the response if given the control is r_{Ci}. The effect on individual i caused by the treatment is a comparison of r_{Ti} and r_{Ci}, such as $r_{Ti} - r_{Ci}$. The challenge in causal inference is that we see r_{Ti} or r_{Ci}, not both.

A causal effect is a comparison of the outcome a particular person would exhibit if given the treatment and the outcome this same individual would exhibit if given the control condition. Causal inference is challenging because, for any individual, we never see both outcomes.

For individual i, treatment T has no effect compared to the control C if individual i would have the same outcome under treatment T and control C, so that $r_{Ti} = r_{Ci}$ and the causal effect is zero, $0 = r_{Ti} - r_{Ci}$. If this is true for everyone—if $r_{Ti} = r_{Ci}$ for $i = 1, 2, \ldots, I$—then treatment T has no effect compared to control C for everyone in the finite population of I people.

In a population of I individuals, $i = 1, 2, \ldots, I$, the average treatment effect is the average of I causal effects, $r_{T1} - r_{C1}, r_{T2} - r_{C2}, \ldots, r_{TI} - r_{CI}$. The ATE is the average of I numbers, but none of these numbers is observed.

Flipping a Fair Coin to Assign Treatments to People

A curious thing happens if we flip a fair coin to decide whether Kim or James is given treatment. A fair coin is like a fair lottery: it ignores Kim and James and their attributes, and comes up heads with a probability of one-half. If the coin toss is heads, Kim receives treatment T, James receives control C, and the estimated effect is $r_{Tk} - r_{Cj} = 1 - 0 = 1$, as above. If the toss is tails, James receives treatment T, Kim receives control C, and the estimated effect is $r_{Tj} - r_{Ck} = 0 - 1 = -1$, as above. The right answer is zero—no effect for Kim or James—but no matter how the coin comes up, we get the wrong answer: 1 for a heads, –1 for a tails. The curious fact is that if we could somehow average over the head and tail, we would get the right answer, as $0 = (1 \times \frac{1}{2}) + (-1 \times \frac{1}{2})$. Is this curious fact useful?

It does not look useful. We can see the world in which the coin comes up heads or the world in which the coin comes up tails, but we cannot see both worlds, so we cannot average over the two worlds with a probability of one-half attached to each world. We do not know whether the coin will come up heads or tails, but we do know that we will get the wrong answer—not zero effect—no matter how the coin comes up. That does not look useful, not at first.

Of all the virtues of human character, the most underrated is persistence. Suppose that we have many pairs of two people, not just Kim and James, and suppose that we flip the coin again and again, once for each pair of two people, assigning treatment T or control C in one pair after another based on the sequence of heads and tails. Suppose, finally, that the treatment has no effect on any of these people, so we are still trying to estimate zero for no effect.

In this case, with many coin flips for many pairs, ask, Does zero equal the average response of individuals given T minus the average response of individuals given C? No, but this difference of averages will be close to zero, unless we obtain an extremely unlucky, an extremely improbable, sequence of coin flips. Now *that* looks useful. Something we can see and calculate from experimental data is, with high probability, close to zero when a treatment has no causal effect. We can only see the one actual world, but we can learn something about possible worlds that we cannot see, and all it takes to see into unrealized possible worlds is repeated flips of a fair coin to assign treatments. Given that this sounds useful, we should pause to understand it.

The premise is that the treatment has no effect, so we are trying to estimate zero. We will estimate other things later, but for now, let us work under this premise and get to the point where we hit zero when the target is at zero. So the premise throughout the several remaining paragraphs of this section is that there is zero treatment effect. If the treatment has no effect on all the individuals in the various pairs, then the two potential outcomes for each individual are equal. That is, for any individual, say individual i, the two potential outcomes are equal, $r_{Ti} = r_{Ci}$, and the causal effect is zero, $0 = r_{Ti} - r_{Ci}$.

If the treatment has no effect, as we are assuming, then the pairs divide into three types, always with $0 = r_{Ti} - r_{Ci}$. In the first type of pair, there are two survivors—two people who would both survive whether given treatment T or control C. For both of these individuals, the two potential outcomes both equal one for survival, $r_{Ti} = r_{Ci} = 1$. In such a pair, no matter who is picked by the coin for T and C, the difference in responses for the treated person and control person in the pair is $1 - 1 = 0$.

In the second type of pair, there are two people fated to die whether they receive T or C. For both of these individuals, the two potential outcomes are both zero, $r_{Ti} = r_{Ci}$

= 0. In the second type of pair, the difference in outcomes between the person given T and the person given C is $0 - 0 = 0$ no matter how the coin turns.

The third type of pair resembles Kim and James: one person lives, the other dies, and switching from T to C does not change that for either person. Here, it is important that the flip of a fair coin assigned one person to treatment T, the other to control C. In such pairs, as with Kim and James, the difference between the outcome of the person given T and that of the person given C is $1 - 0 = 1$ with a probability of one-half or $0 - 1 = -1$ with a probability of one-half. Over many such pairs, the average of the ones and minus ones tends to zero because of a theorem in probability theory called the law of large numbers. Half the time, we get a one. Half the time, we get a minus one. Do that over and over, and the average over a large number of coin flips cannot be far from zero.

Now consider the proportion of survivors among individuals randomly picked for T compared to those randomly picked for C. Pairs of types one and two contribute exactly zero to the difference in proportions, no matter how the coin falls. Pairs of type three contribute a random quantity, a 1 half the time and a −1 half the time, depending on how the coin falls. So the difference in the total number of survivors under T and C is made up of lots of zeros, for pairs of types one and two, and lots of random ones or minus ones that depend on how the coin falls in each new flip of the coin. Divide that total by the number of pairs—that is, the number of coin flips—to get the difference in the proportions of survivors under T and C. The difference in the proportions gets closer to zero as the number of pairs increases because pair differences of one are canceling pair differences of minus one. The cancellation is imperfect because the coin flips are random, but as the number of flips increases, the cancellation in the difference of proportions approaches perfection owing to the law of large numbers. So the difference in the proportion of survivors in groups T and C tends to zero when there is no treatment effect.

This argument took several paragraphs, so let us recap. With one pair, Kim and James, we considered flipping one coin to assign one of them to treatment T and the other to control C; that was correct if we averaged over heads and tails, but it was entirely wrong no matter how the coin fell. But we persisted. What did not work for one pair and one coin flip did work if we did it many times, with many pairs of two people and many coin flips. With one coin flip, being right on average meant nothing because the answer under both heads and tails was wrong. With many coin flips, being right on average meant everything because being right on average meant that our errors tended to cancel, to average out to zero error, as the number of coin flips increased.

There is nothing mysterious or magical about this distinction between one and many coin flips. The distinction has something to do with coins and nothing in particular to do with causal inference. It is one of the distinctions between being a gambler and a casino. The gambler gets something noisy and unstable, sometimes wins and sometimes loses, sometimes goes home happy and sometimes sad. The casino wins some bets and loses others, but ultimately gets the average. Casinos never gamble; they collect the average, always. The art of being a casino is to offer everyone slightly unfair bets that favor the casino and then collect the average. Being a casino is boring in a pleasant way; the casino always wins, at least on average, and with many coin flips the average is all there is. The magic is not in the coin flips. The magic is in reducing causal inference to coin flips, peering into possible worlds that never happened by flipping coins.

So given a fair coin flipped repeatedly, an elementary causal inference is possible and even routine. With sufficiently many coin flips for sufficiently many people, no effect will look like no effect. The fair coin is essential; remove it, and the argument falls apart. If someone looks at Kim and says, "You look like the type of person who will benefit from treatment T," and then turns to James and says, "I don't like your looks, so you are getting control C," then all bets are off. Assign treatments inequitably, and you could be mistaken about the effects of those treatments for twenty centuries, as we have seen.

In brief, causal effects compare potential outcomes in possible worlds, but we see only the one actual world. At first, that seems to be a problem. How can we learn about possible worlds that did not occur from the one actual world that did occur? Then we did something quite radical: we let coin flips determine the actual world. For many pairs of people, we let coin flips decide who was treated and who was the control in each pair. This radical step means that the world we actually saw—the real world—was picked at random from many possible worlds. Certain averages in the actual world are then pulled toward certain averages over possible worlds by the properties of the coin flips, in particular by the law of large numbers applied to coin flips. In short, there are aspects of those many possible worlds that we can infer from the one actual world, because the actual world was actually built by flipping a coin again and again. This topic is explored in chapter 2.

The Average Treatment Effect (ATE)

If we flip coins to assign treatments, then we can recognize a treatment that has no effect. What about treatments that do have an effect?

Whether or not the treatment affects Kim or James, the average effect of the treatment on Kim and James is the effect for Kim, $r_{Tk}-r_{Ck}$, plus the effect for James, $r_{Tj}-r_{Cj}$, divided by two, namely ATE = $(1/2)$ $(r_{Tk}-r_{Ck} + r_{Tj}-r_{Cj})$. If the treatment has no effect on either Kim or James, then ATE

= 0. If ATE $=1/2$, then giving T rather than C would save the life of either Kim or James, but would not alter the survival of the other. As before, the ATE cannot be calculated because we see either r_{Tk} or r_{Ck} for Kim but not both, and either r_{Tj} or r_{Cj} for James but not both.

If we gave everyone the treatment—both Kim and James—then we would observe the average response of everyone to the treatment, namely $r_{T+} = (r_{Tk}+r_{Tj})/2$. Instead, if we gave everyone the control, then we would observe the average response to the control, $r_{C+} = (r_{Ck}+r_{Cj})/2$. We could see r_{T+} if everyone received the treatment, or we could see r_{C+} if everyone received the control, but we cannot see both r_{T+} and r_{C+} for the same reason that we cannot see the ATE. Rearranging these expressions shows that $r_{T+}-r_{C+} =$ ATE.[6] This is the same algebraic rearrangement that says for thirty students in one class, the average of the thirty differences in their individual grades on the midterm and the final may be computed as the difference between the average of thirty grades on the midterm minus the average of thirty grades on the final; it does not matter whether you subtract first and then average, or average first and then subtract.

Algebraic rearrangement demonstrates the equality of two quantities that we cannot see, $r_{T+}-r_{C+} =$ ATE. Is that useful? It does not look useful, not at first.

RANDOMIZED EXPERIMENTS

The Effects of Treatments for Ebola Virus Disease

The first symptoms of Ebola virus disease resemble symptoms of flu: fever, fatigue, headache, muscle pain, and sore throat. Then comes vomiting, diarrhea, decreased liver and kidney function, internal and external bleeding, bleeding gums, and blood in the stool. Half of those infected may die.[1]

Sabue Mulangu, Lori Dodd, Richard Davey Jr., and their colleagues conducted a randomized clinical trial comparing several treatments for Ebola virus disease. A joint effort of Kinshasa University and the US National Institute of Allergy and Infectious Diseases, the trial was conducted in the Democratic Republic of the Congo between November 2018 and August 2019. It was called the PALM trial, an acronym derived from a phrase in the

Kiswahili language meaning "together save lives," an apt phrase describing randomized clinical trials in general. My discussion of this trial simplifies some of its features.[2]

The memorably named PALM trial compared two drugs with the less memorable names ZMapp and mAb114. The first drug, ZMapp, was engineered from several monoclonal antibodies derived from immunized mice; it appeared to work in experiments on nonhuman primates. The second drug, mAb114, was derived from monoclonal antibodies obtained from one human survivor of an Ebola outbreak. It too appeared to work for nonhuman primates. Which drug is better in humans? Which drug saves more lives?

The real heroes of a battle are the soldiers who fought it. The real heroes of a clinical trial are the patients who participated. Are patients safer inside or outside a trial? We like to imagine that all health care providers make our health their overriding priority, with financial interests a distant second. Whatever we like to imagine about non-experimental health care, there are certainly more people keeping an eye on what is going on inside a clinical trial sponsored by the US National Institutes of Health. In the PALM trial, the treatment plans and study design were reviewed by two ethics committees, one at Kinshasa University, and the other at the US National Institute of Allergy and Infectious Diseases. To participate in the trial, a patient (or parent) needed to give written informed consent. An independent committee, or "data and safety monitoring board," kept an eye on the investigators and patient outcomes. In the PALM trial, the data and safety monitoring board terminated early two of what turned out to be the less successful treatments based on a carefully planned interim analysis, so no additional patients received those treatments.

What did these two ethics committees actually do? Among other things, they checked for equipoise among the treatments being compared. Alex London wrote,

> The principle of equipoise states that if there is uncertainty or conflicting expert opinion about the relative therapeutic, prophylactic, or diagnostic merits of a set of interventions, then it is permissible to allocate a participant to receive an intervention from this set, so long as there is not consensus that an alternative intervention would better advance that participant's interests. . . . [T]he presence of equipoise ensures that each participant receives an intervention that would be recommended or utilized by at least a reasonable minority of informed expert clinicians.[3]

John Gilbert, Richard Light, and Frederick Mosteller noted,

When we object to controlled field trials with people, we need to consider the alternatives—what society actually does. . . . [W]e spend our money, often put people at risk, and learn very little. This haphazard approach is not "experimenting" with people; instead, it is fooling around with people.[4]

The contrast between experimenting with human subjects and fooling around with human subjects is the contrast between the PALM trial and bleeding George Washington.

A Randomized Controlled Trial

The PALM trial treated 169 patients with ZMapp and 174 patients with mAb114. How were patients selected for one drug or the other? The assignment of patients to the drugs was done by a truly random device—in effect, by flipping a fair coin. In the state of equipoise—in ignorance of the better treatment—the trial was equitable: each patient had the same chance of receiving the better treatment, whatever that was. The ZMapp drug was discontinued early when it became evident that it was the inferior treatment. The trial favored no one: no attribute of a person predicted the treatment that person received; receiving mAb114 rather than ZMapp was purely good luck. The inferior treatment, ZMapp, was discarded in under ten months, not more than twenty centuries, and the superior treatment, mAb114, was given to half the patients in the PALM trial before it was known to be superior.

Why Assign Treatments at Random?

There are several closely connected reasons. Consider, first, the simplest reason, though not the most important one. In the published report about the PALM trial, the first table—the so-called balance table—compared the two groups of patients receiving either ZMapp or mAb114 in terms of their condition prior to treatment. A quantity describing an individual prior to treatment is called a covariate, so the balance table describes covariates. For instance, a person's age at the start of treatment is a covariate because receiving ZMapp as opposed to mAb114 does not alter age. Because a covariate describes an individual prior to treatment, we know that its value is unaffected by a treatment the individual has not yet received.

What did the balance table reveal? As one might expect in a fair lottery with 174 winning mAb114 tickets and 169 losing ZMapp tickets, the winners and losers were not very different before treatment. After all, any differences prior to treatment were due to chance, the coin flips that assigned one patient to ZMapp and another to mAb114. The study included adults, children, and even a few neonates under seven days old, yet the mean age in the two treatment groups was similar: 29.7 years in the ZMapp group and 27.4 years in the mAb114 group. The groups were also similar in terms of sex and weight, vaccination history, measures of the nature and intensity of the current illness, and blood chemistry and vital signs. The published table compared a total of twenty-seven baseline measures, finding again and again that the groups were reasonably similar prior to treatment.

Consider sex. In the ZMapp group, 87 of the 169 patients were female, or 51.5 percent. In the mAb114 group, 98 of the 174 patients were female, or 56.3 percent. Sex is a covariate; being assigned to ZMapp rather than mAb114 did not alter anyone's sex. So the 51.5 versus 56.3 percent difference in the percent of females is due to chance, the way fair coins fall. We can say this more precisely. We could certainly flip the $343 = 169 + 174$ coins again, producing a new ZMapp group and new mAb114 group, and this new random assignment would have a certain percentage of females in each group. With a computer, we could do this many times, millions of times. If we built millions of randomized experiments by flipping 343 coins for each of millions of experiments, then we would know exactly what coin flips do to the percentages of females in the ZMapp and mAb114 groups. The actual PALM trial is entirely typical of these millions of experiments: 39 percent of these millions of experiments produce a bigger difference in the percent of females than actually occurred in the PALM trial, and 61 percent produce a smaller difference. The 51.5 versus 56.3 percent difference in the proportion of females would tend to be smaller if the trial were larger, again because of the law of large numbers. Clinical trials commonly report balance tables, perhaps to brag about their success at balancing covariates or perhaps to demonstrate that the investigators did not mess things up along the way; we know, however, what coin flips do to covariates without looking at balance tables. Coin flips can be studied mathematically, so we do not need to run the computer to know what fair coins do.

If our goal were to compare two groups of patients with the same distribution of ages, then we could produce groups with closer mean ages than 29.7 and 27.4; however, that is not the goal. To produce closer mean ages, we would have to use age when assigning treatments. We would have to say, "The mean age so far in the ZMapp group is a bit too high, so the next patient over 40 should be assigned to mAb114, and the next patient under 20 should be assigned to ZMapp." The problem is that we can do this for age or other covariates that we measured, but debates about cause and effect invariably refer to some covariate that was not measured. "True, the groups look comparable in the balance table," a critic of the study concedes, "but appearances deceive, for the groups differ in terms of a specific covariate that was not measured." Perhaps the critic mentions some genetic variant that the investigators did not measure or speculates about such a variant that will be discovered a decade from now.

Because they used coin flips to assign treatments, the investigators in the PALM trial have a strong rebuttal to the concerns raised by this critic. They would say, "There is every reason to believe, and no reason to doubt, that the two treatment groups are comparable in terms of this genetic variant that you speculate will be discovered a decade from now. We showed you the balance table for observed covariates to reassure you, but we had no part in creating the comparability you see there. What you see in the balance table is what fair coins do. Those coins balanced age, sex, and blood chemistry, but the coins knew nothing of age, sex, and blood chemistry. So those same coins likely balanced your genetic variant as well."

This strong rebuttal to concerns about unmeasured covariates is a second and more important reason for randomized treatment assignment. Anyone can design an experiment that balances an observed covariate like age. It takes some doing to design an experiment to balance a covariate that will be discovered ten years from now. Randomization balances both covariates. It is a convenience that randomization tends to balance measured covariates, but it is a small miracle that it balances unmeasured covariates and those that will not be discovered for decades to come.

It is a convenience that randomization tends to balance measured covariates, but it is a small miracle that it balances unmeasured covariates and those that will not be discovered for decades to come.

Randomized Treatment Assignment and Causal Inference

There is a third and most important reason for randomized treatment assignment. Randomization is a basis for inference about causal effects, as discussed in chapter 1. The theory of causal inference in randomized experiments was developed in the 1920s by Sir Ronald Fisher. Two aspects of this theory are explored in the two sections that follow: the estimation of the average treatment effect and testing the hypothesis of no causal effect.

Randomization does not just balance covariates. It also balances the potential outcomes (r_{Ti}, r_{Ci}), $i = 1, 2, \ldots, I$, that define causal effects. The coin flips care nothing about person i and assign people fairly no matter who they are. In particular, the coin flips care nothing about the fate of person i under treatment or control—they care nothing about (r_{Ti}, r_{Ci})—so they tend to balance (r_{Ti}, r_{Ci}) in treated and control groups. This turns out to be the reason that the difference in the mean responses in the treated and control groups estimates the average treatment effect (ATE).

Estimating the Average Treatment Effect

The effect on individual i caused by a treatment is a comparison of the response, r_{Ti}, that this individual would exhibit if given the treatment, T, and the response, r_{Ci}, that this same individual would exhibit if given the control, C. In the PALM trial, the primary outcome was survival to twenty-eight days after the start of treatment: 1 for survived, 0 for died. Referring to mAb114 as treatment T and ZMapp as treatment C, each patient has a potential survival, r_{Ti}, if given mAb114, and a potential survival, r_{Ci}, if given ZMapp. In particular, mAb114 would save the life of person i if $r_{Ti} = 1$ and $r_{Ci} = 0$. Our problem is that we can see either r_{Ti} or r_{Ci}, not both, for any one individual i. As with Washington, for any one individual i, we can only speculate about that individual's survival under a treatment that individual did not receive. For a population of individuals, experimentation replaces speculation.

For the $I = 343$ patients receiving either mAb114 or ZMapp, the average treatment effect is the average of 343 effects, $r_{Ti}-r_{Ci}$, for individuals $i = 1, 2, \ldots, 343$, and none of the 343 effects were observed. We cannot calculate the average treatment effect from the data in the PALM trial; what we need for that calculation is not there.

At the end of chapter 1, routine algebra showed that the average treatment effect equals the difference of two other quantities that we cannot calculate. Specifically, $\text{ATE} = r_{T+}-r_{C+}$, where r_{T+} is the proportion of the $I = 343$ patients who would have survived had they all been given mAb114, and r_{C+} is the proportion of the $I = 343$ patients who would have survived had they all been given ZMapp.

We cannot see r_{T+} because only 174 of the 343 patients received mAb114, and we cannot see r_{C+} because only 169 of the 343 patients received ZMapp. Routine algebra is not enough for causal inference.

Nonetheless, the PALM trial has what it takes for causal inference. Although we did not see the rate of survival under treatment mAb114 for all $I = 343$ patients, we did see that rate for roughly a random half of the 343 patients, or specifically for 174 patients. Of the 174 patients who received mAb114, 113 survived to twenty-eight days and 61 died, so the proportion who survived is 113/174 = 64.9 percent. This sample proportion is not r_{T+}, because r_{T+} would be calculated from all $I = 343$ patients, but intuition suggests it is a good estimate of r_{T+}. I will discuss why that intuition is sound in a moment. In parallel, of the 169 patients who received ZMapp, 85 survived to twenty-eight days and 84 died, so the proportion who survived is 85/169 = 50.3 percent. Again, that sample proportion is not r_{C+}, because r_{C+} depends on all 343 patients, but the sample proportion is a credible estimate of r_{C+}. The average treatment effect is ATE = $r_{T+} - r_{C+}$, so a reasonable estimate of the ATE is 0.649–0.503 = 0.146, or an increase of 14.6 percent surviving if given mAb114 instead of ZMapp.

In what sense is 113/174 = 64.9 percent a reasonable estimate of the survival rate, r_{T+}, had everyone—all 343 patients—been given mAb114? It feels comforting that we saw the survival under mAb114 of slightly more than half of the 343 patients, but in truth we should not be comforted by this. Had mAb114 been given to the 174 youngest patients, the 174 patients with the lowest fevers, or the 174 patients with the lowest viral loads, then $113/174 = 64.9$ percent would have been a bad estimate of what would have happened to all 343 patients had they all been given mAb114. The reason $113/174 = 64.9$ percent is a reasonable estimate of the survival rate, r_{T+}, under mAb114 is that the 174 patients were picked at random for mAb114. Giving mAb114 to the 174 patients with the lowest fevers is unimaginably awful when compared to random selection. There are 7.4×10^{101} ways to pick 174 of 343 people to receive mAb114, and picking the 174 patients with the lowest fevers is the single worst way to pick so far as imbalances in fevers are concerned. It is inconceivable that random selection would make this awful selection. Randomization is quite likely to balance fevers, as indeed it did, with pretreatment temperatures of 37.4 degrees centigrade in the mAb114 group and 37.5 degrees centigrade in the ZMapp group.

Much more important than balancing fevers, randomization is likely to balance the potential survival, r_{Ti}, under mAb114. It is precisely this that makes $113/174 = 64.9$ percent a good estimate of the proportion who would survive if all 343 patients had received mAb114. With a covariate like pretreatment temperature, we see the values in both groups, 37.4 degrees in the mAb114 group and

37.5 degrees in the ZMapp group, but had the temperature been recorded only in the mAb114 group, we would not have missed much. In parallel, we have not missed much about the survival rate, r_{T+}, for all 343 patients under mAb114 by observing survival, r_{Ti}, under mAb114 only for a random 174 patients.

Chapter 1 considered flipping a coin to assign either Kim or James to the treatment, and the other to the control. The estimate of the average treatment effect was Kim under the treatment minus James under the control if the coin came up heads, $r_{Tk}-r_{Cj}$, or James under the treatment minus Kim under the control if the coin came up tails, $r_{Tj}-r_{Ck}$. The estimate was far off the mark whether the coin came up heads or tails. There was, however, a curious fact. If somehow we could do what we cannot do, namely average over the two cases, heads or tails, with equal probabilities ½, then the average over heads and tails was equal to the ATE. Averaging over coin flips in this way is calculating the so-called expectation of the estimate of the ATE, so we found that the expectation of the estimate equaled the quantity we wanted to estimate, namely the ATE. An estimate with this property is said to be "unbiased," a technical term. So the Kim versus James comparison was unbiased as an estimate of the ATE, but it was so unstable, so dependent on the head or tail, as to be useless.

The PALM trial preserved the strength and removed the weakness of the Kim versus James comparison. In the

PALM trial, the estimate of the ATE was unbiased, as it was when we flipped a coin to assign Kim or James to treatment, but in the PALM trial the estimate is more stable. Obtaining an unbiased estimate was the hard part—that required random assignment of treatments—but stabilizing the estimate merely requires more people and more coin flips.

In precise terms, what does that mean? The estimate of the ATE, namely $0.649 - 0.503 = 0.146$, is unbiased. That is, if we averaged over the 7.4×10^{101} ways to pick 174 of 343 people for mAb114, giving equal probability to each way, then the difference in the proportions of survivors in the treated and control groups would average out to equal the ATE.[5] The difference between flipping one coin for Kim and James and flipping 343 coins in the PALM trial is the difference between being a gambler and being a casino. In chapter 1, we imagined Kim survived under both treatments and James died under both treatments, so the treatments did not differ in their effect. In this case, for Kim and James, the estimate of the ATE would change from 100 percent caused to survive to 100 percent caused to die as the coin changed from heads to tails. Because of the larger sample size in the PALM trial, the difference in the sample proportions surviving in the mAb114 and ZMapp groups is more stable. In the PALM trial, different coin flips would pick different people for mAb114, and so produce different estimates of the ATE, but there is only a small chance that 343 coin flips will produce an estimate that is far from the ATE.

The story is told of a mathematician who was asked how he would move a book from the table to the floor. He answered, "I would pick up the book, bend over, and when the book is in close proximity to the floor, I would release it." He was then asked how he would move a book from the chair to the floor. He replied, "I would bend over, pick up the book from the chair, stand up straight, and place the book on the table, thereby reducing this new problem to one I have previously solved."

Following this mathematician's sound advice, randomized treatment assignment reduces an apparently insoluble problem to a routine technical one that had been previously solved. In chapter 1, the apparently insoluble problem is to peer into possible worlds that never happened and compare those unrealized possible worlds to what did happen in the actual one. Causal effects are comparisons of what did happen with what would have happened if people had received different treatments. Randomized treatment assignment has reduced this problem to the minor technical problem of drawing an inference about a finite population of people on the basis of a probability sample from that population.[6] Expressed differently, if we design an experiment so that the actual world is a random draw from a set of possible worlds, then we can draw inferences about certain aspects of worlds that were never realized. For instance, we can estimate the average effect of the treatment with negligible error in a sufficiently large randomized trial.

Testing the Hypothesis of No Effect

The hypothesis of no causal effect asserts that some people live and others die, but receiving mAb114 rather than ZMapp has nothing to do with it. Of the 174 patients who received mAb114, 113/174 = 64.9 percent survived for twenty-eight days, and of the 169 patients who received ZMapp, 85/169 = 50.3 percent survived. Is that solid evidence that the hypothesis of no effect is false? Could the difference be due to chance—an unlucky sequence of coin flips assigning people to mAb114 or ZMapp—as opposed to an effect caused by the treatments?

In the PALM trial, we saw several differences that we know are due to chance, the coin falling heads or tails. The coin flips put 87 women and 82 men in the ZMapp group, for 87/(87 + 82) = 51.5 percent female in the ZMapp group. The coin flips put 98 women and 76 men in the mAb114 group, so the group was 98/(98 + 76) = 56.3 percent female in the mAb114 group. That difference, 4.8 percent = 56.3 percent − 51.5 percent, is due to chance, the idiosyncratic way the coin fell time and again. Could the difference in survival, 14.6 percent = 64.9 percent − 50.3 percent, also be due to chance, not a treatment effect?

Obviously the difference in survival *could* be due to chance providing we accept anything that is logically possible as realistically possible. No one does that, of course; you could not cross the street if you thought that way. Many things that are logically possible are ridiculously improbable.

In the population of 343 patients, there were 113 + 85 = 198 survivors or 198/343 = 57.7 percent. In a population of 343 patients with 198 survivors, if there were no treatment effect, it is logically possible that a sequence of coin flips will pick 174 patients for mAb114 with 113 survivors and also pick 169 patients for ZMapp with 85 survivors. There are, in fact, many ways this might happen. Indeed, there are 1.51×10^{99} different sequences of 343 heads and tails that do precisely this. Although 1.5×10^{99} at first appears to be an impressively large number, there are 7.4×10^{101} sequences of 343 coin flips that pick 174 patients for mAb114. Could the difference in survival, 14.6 percent = 64.9 percent − 50.3 percent, be due to chance? There is clearly more to this question than logical possibility.

What precisely is the hypothesis that says mAb114 and ZMapp do not differ in their effects? To speak concisely, we often speak of the hypothesis of no effect, but we mean that the treated and control conditions do not differ in their effects. This hypothesis says that each patient *i* of the 343 patients would have the same survival to twenty-eight days whether given mAb114 or ZMapp. As discussed in chapter 1, writing T for mAb114 and C for ZMapp, the hypothesis of no effect claims that $r_{Ti} = r_{Ci}$ for each person i, for $i = 1, 2, \ldots, 343$. This hypothesis is commonly called "Fisher's hypothesis of no effect" because it played an important role in his theory of randomized experimentation. What would you have to believe about the coin flips to believe this hypothesis? If this hypothesis were true, would a difference in survival of 14.6 percent = 64.9 percent − 50.3 percent be a common event, like getting two heads in two flips of a fair coin, or a fairly rare event, like flipping a coin seven times and getting seven heads?

The chance of two heads in two flips of a fair coin is $1/2^2 = 1/4$, but the chance of seven heads in seven flips of a fair coin is $1/2^7 = 0.0078$. If 1,000 people flip a fair coin twice, we expect 250 people to get two heads. If 1,000 people flip a fair coin seven times, we expect fewer than 8 people to get seven heads. Flipping a coin seven times and getting seven heads would be reason to doubt the coin is fair, but flipping a coin twice and getting two heads is no reason to doubt the coin is fair.

Does survival in the PALM trial resemble two heads in two flips or seven heads in seven flips? If there is no difference between mAb114 and ZMapp among all 343 patients, is a difference in survival of 14.6 percent a rare or common event?

At this point, two details require attention. In fact, mAb114 beat ZMapp by 14.6 percent. If mAb114 had beat

ZMapp by more than 14.6 percent, then we would have been even more impressed. So we really want the probability of a difference as large or larger than 14.6 percent, not the probability of exactly a 14.6 percent difference. Also, we did not know before the trial that mAb114 would be the winner. Had ZMapp won by 14.6 percent, then we would be talking about the chance that ZMapp won by at least 14.6 percent.

Fixing these two details, the revised question is, If there were no difference in the effects of mAb114 and ZMapp, then what is the probability that coin flips alone would produce an apparent difference in survival, positive or negative, as large or larger than the difference we actually saw? The answer, it turns out, is 0.0083. The difference in survival in the PALM trial is much closer to seven heads in seven flips of a fair coin, with its probability of $1/2^7 = 0.0078$, than it is to two heads in two flips, with its probability of $1/2^2 = 0.25$. It would take quite an unlikely sequence of coin flips to produce a difference in survival of 14.6 percent = 64.9 percent − 50.3 percent in the absence of genuine reduction in mortality caused by mAb114.

Where does that probability, 0.0083, come from? It comes from the behavior of fair coin flips. We could turn the task over to the computer. The hypothesis of no effect says that the population of 343 patients contains 113 + 85 = 198 people who would survive whether given mAb114 or ZMapp, 145 = 343 − 198 people who die whether given mAb114 or ZMapp, and no one whose survival would change if one drug were substituted for the other. The hypothesis may be true or false, but that is what the hypothesis says. In this hypothesized population, we could tell the computer to flip a fair coin 343 times to assign patients to mAb114 or ZMapp. That would produce some mAb114 minus ZMapp difference in survival rates. That difference in survival would be due to chance because in the hypothesized population, no one's survival depends on which drug is given. We could tell the computer to do that calculation over and over. If the computer did the task 1,000 times, creating 1,000 phony PALM trials, we expect about 8 of the 1,000 to produce a difference in survival as large or larger than the one we saw, and about 992 of the 1,000 to produce a smaller difference. The difference in survival that we saw, 14.6 percent = 64.9 percent − 50.3 percent, *could* be due to chance—it is a logical possibility—but it is quite an improbable possibility.

To recap briefly, the reasoning is as follows. We asked, If there were no difference between mAb114 and ZMapp, could the observed difference in survival rates of 14.6 percent = 64.9 percent − 50.3 percent be due to an unlucky sequence of coin flips assigning one patient to mAb114 and another patient to ZMapp? We found that it is logically possible but quite improbable: the probability of such a sequence of coin flips is 0.0083. To maintain that mAb114 and ZMapp do not differ in their effects on survival, you must maintain that you accidently observed a quite improbable sequence of coin flips.

What Is Special about Coin Flips?

Randomized experiments assign treatment or control on the basis of a new flip of a fair coin. What property of coin flips is important in randomized experiments? What properties are incidental?

It is not important that the coin comes up heads half the time. We could roll a die and assign an individual to treatment when a one or two appears, and assign control when a three, four, five, or six appears. In this situation, the probability of treatment is one-third and the probability of control is two-thirds, but it is still a completely randomized experiment. Experiments of this sort are sometimes performed when treating an individual is expensive and the control condition is inexpensive.

Coin flips and die rolls share a crucial property: they produce a fair lottery. The probability of a winning ticket in this lottery is not critical. What is critical is that everyone has the same chance of a winning ticket. With a coin, the chance of a winning ticket is half, but it is half for everyone. With the die roll in the previous paragraph, the chance of a winning ticket is one-third, but it is one-third for everyone.

Coin flips and die rolls share a crucial property: they produce a fair lottery. What is critical is that everyone has the same chance of a winning ticket.

Each of us is unique. It is impossible to assign unique people to treated and control groups so that the groups are the same. Washington was unique, and the treated and control groups cannot be identical if he is in one group and not the other. Randomization does not make unique people the same; that cannot be done. By virtue of being a fair lottery, randomization makes receiving treatment or control unrelated to everything that makes people different. Before it happens, we often remark, This is the type of person who goes to college, ends up in prison, or will be a good father. In contrast, in a randomized experiment, before it happens, we can never say, This is the type of person who ends up in the treated group. Because it is a fair lottery, there is no type of person who ends up in the treated group. Make up types of people in any way you like, but a person's type will never predict their treatment, for it will never predict the turn of a fair coin.

Bleeding George Washington and the Theory of Humors

Could eighteenth-century physicians have discovered that they harmed patients by bleeding them? Think about physicians during that time. They had coins. They knew how to flip them. They could measure the outcome, distinguish the dead from the living. They even had a basic understanding of probability. What did they lack? Perhaps what Dewey called, as mentioned earlier, the experimental habit of mind.

Had the eighteenth century discovered that patients were harmed by bleeding, physicians might have questioned the structure of medical knowledge built on the theory of humors. The success or failure of treatments in randomized trials might have stimulated basic research in the biology of disease; this in turn might have produced better treatments to evaluate in further trials—as it does today.

OBSERVATIONAL STUDIES

The Problem

What Are Observational Studies?

Randomized experimentation is unethical or impractical in some contexts. A harmful treatment cannot be imposed on a person. You cannot force someone to smoke cigarettes to discover the diseases caused by smoking. You cannot inflict emotional trauma to understand posttraumatic stress syndrome. The US Environmental Protection Agency may need to set a standard for exposure to a potential toxin or carcinogen, such as radon gas, but the agency cannot conduct a randomized experiment on humans to determine the extent of harm, if any, caused by such an exposure. Instead, it might use evidence from various sources, including nonrandomized or observational studies of human exposures, such as the exposure of uranium miners to radon gas.[1] The decisions that individuals make about education, consumption of alcohol and narcotics, and savings and debt affect their prosperity, as do government decisions about public and private education and minimum wages, but you cannot study these effects in randomized trials.

In 1964, the US Surgeon General published a report, *Smoking and Health*, concluding on the basis of nonrandomized studies that smoking causes lung cancer. The following year, a member of the advisory committee that prepared the report, William Cochran, defined an observational study as an empiric investigation in which the "objective is to elucidate cause-and-effect relationships [in which it] is not feasible to use controlled experimentation, in the sense of being able to impose the procedures or treatments whose effects it is desired to discover, or to assign subjects at random to different procedures."[2]

An observational study has the same goal as a randomized experiment—the estimation of the effects caused by treatments—but treatments are not assigned by coin flips or random numbers from the computer. Perhaps the individuals chose their own treatments, as happens with smoking cigarettes. Perhaps the minimum wage is increased in one US state and held steady in another.[3] Perhaps a natural disaster, such as an earthquake, abruptly inflicts emotional trauma in one region while sparing other regions.[4] Perhaps the US Supreme Court strikes down a law as unconstitutional, suddenly altering a government policy.[5] What can be learned from observational studies? What problems arise because treatments are not randomly assigned?

Smoking and Periodontal Disease

Periodontal disease involves the gradual separation of teeth and gums. Does cigarette smoking cause periodontal disease? Scott Tomar and Samira Asma asked this question using data from the US National Health and Nutrition Examination Survey III (NHANES), concluding that the answer is yes, and the US Centers for Disease Control reached the same conclusion. To illustrate issues arising in observational studies, let us do a similar comparison using the more recent 2011–2012 NHANES, which also measured periodontal disease.[6]

Daily smokers are compared to controls who never smoked. Daily smokers smoked every day for the last thirty days, have smoked for more than 30 years on average, and 90 percent have smoked for more than 14.9 years. Controls smoked fewer than a hundred cigarettes in their lives. The comparison involves 441 daily smokers and 1,506 controls whose teeth were examined.

Is it safe to compare smokers and nonsmokers given that people decided for themselves whether to smoke? Are smokers and nonsmokers comparable? When people decide for themselves whether to smoke, do they appear to do so at random, like coin flips or die rolls? Does this observational study look like a randomized experiment?

Actually, the smokers and controls were quite different. Most smokers were men (177 women and 264 men), but most nonsmokers were women (901 women and 605 men). As we will see, a better way to say this is that 16.4 percent of the women smoked, but 30.4 percent of the men smoked.

From sex alone, self-selected smoking behavior looks nothing like a randomized experiment. There were 441 smokers among 1,947 individuals, so 22.7 percent are smokers. In concept, it is easy to pick people at random so that each person has a 22.7 percent chance—a 0.227 probability—of being a smoker. Obtain 1,947 tickets, mark 441 with "smoker" and 1,506 with "control." Put the tickets in a barrel and shake thoroughly. Then reach in and grab a ticket. That ticket assigns the first person. Put the ticket back in the barrel, shake again, and grab a ticket again to assign the second person. And so on, 441 times.

The ticket-in-a-barrel randomization would assign roughly 22.7 percent of people to smoking, but it would be a fair lottery. As a fair lottery, it would be unlikely that 16.4 percent of the women and 30.4 percent of the men would get winning tickets. That does not sound fair at all. In fact, because we are not talking about the actual data but instead about what would happen in a randomized experiment, we can go through the same reasoning in chapter 2 to determine the chance of such an unfair allocation of winning tickets in a ticket-in-a-barrel lottery with 1,947 tickets. The chance that our fair lottery would produce such an unfair allocation of tickets is miniscule; the probability is 3.2×10^{-13}. Imagine that you did the whole lottery 3.2×10^{13} times—a task that would take many lifetimes. Then you would expect to see such an unfair allocation of tickets only once by chance in 3.2×10^{13} lotteries. This just says that women and men genuinely behave differently; that is, their behavior looks nothing like a randomized experiment. What is worse, women and men are not the only groups of people who behave differently.

Smokers had less education. Only 7.1 percent of individuals who had at least a four-year college degree were smokers, but 29.9 percent of those without such a degree were smokers.

Smokers had less income. Among people with a family income of at least three times the poverty level, 12.1 percent were smokers, but among people with income below three times the poverty level, 29.8 percent were smokers.

Smokers were a bit younger than the controls—an important issue for periodontal disease. The difference in smoking was most pronounced at older ages. Among individuals under sixty years of age, 25.2 percent were smokers, but among those sixty and above, 16.6 percent were smokers.

Looking at attributes one at a time, smokers and non-smokers look quite different. The situation looks much worse if you look at several attributes at the same time. Among women at least sixty with at least four years of college, only 4.3 percent were smokers. Among men under sixty without four years of college, 42.3 percent were smokers. That is nearly a tenfold difference. A tenfold difference in the probability of a winning ticket is nothing like a fair lottery.

This nearly tenfold difference came from just three attributes, with age and education coarsely cut into two embarrassingly broad categories. Some people had less than a ninth grade education. Ages ranged from thirty to over eighty. Some people had no income at all, and others had incomes in excess of five times the poverty level. To preserve confidentiality, NHANES caps the reporting of age at eighty and the reporting of income at five times the poverty level.

What would happen if we used actual age, not age above or below sixty? What would happen if we used actual income, not income above or below three times the poverty level? Even with just these five covariates, the data become too thin to calculate proportions. Cut age, income, and education into five categories each, make race Black or other, and add gender, and there are $5 \times 5 \times 5 \times 2 \times 2 = 500$ categories, but there are only 441 smokers to distribute among these 500 categories.

With a bit of statistical magic, we can estimate a probability of smoking for all 500 categories. Once we are into magic, we do not need categories at all. For each of the 441 smokers and 1,506 nonsmokers, we can estimate a personal probability of smoking for a person with an exactly specified age, income, education, race, and gender. These probabilities are mostly different. The $441 + 1506 = 1,947$ people have 1,753 distinct probabilities. Even with five covariates, few people look identical. In fact, the magic is not mysterious, but it is slightly technical.[7]

The estimated probabilities range from 3.2 to 64.5 percent. That is no longer a tenfold difference; it is more than a twentyfold difference. The lowest estimate, a 3.2 percent chance of smoking, was for a sixty-one-year-old woman with four years of college and an income above five times the poverty level. The highest estimate, a 64.5 percent chance of smoking, was for a fifty-eight-year-old man with less than a ninth grade education and an income that was below the poverty level—about two-thirds of the poverty level.

Before continuing, we need to pause to consider how we can closely inspect $441 + 1,506 = 1,947$ numbers.

Interlude: Tukey's Boxplot

John Tukey proposed the boxplot to quickly see what is important in a large batch of numbers. Once we can do that, we can split the batch in interesting ways—smokers and nonsmokers, say—and quickly see how groups differ.

With lots of data, we need to know the typical value. A typical age among the 1,947 people under study is fifty. The median is fifty years: ignoring a few ties, half the people are under fifty and half are over fifty. The median is one good typical value.

Of course, most people are not fifty years old. Among the 1,947 people, only fifty-four are exactly fifty years old. People vary. Most of us are not typical. The typical person deviates from what is typical. We need to see how people typically vary.

For people above the median, older than fifty years, we ask, Typically, how old are they? The median worked once. Why not try it again? Take everyone older than the median and calculate their median age. That turns out to be sixty-one. Among people older than the median, half are older than sixty-one. In general, that number is called the upper quartile because 25 percent of people are older than sixty-one and 75 percent are younger than sixty-one.

The median worked twice. Why not try it again? Take everyone younger than the median and calculate their median age. That turns out to be thirty-nine. In general, that number is the lower quartile because 25 percent of people are younger than thirty-nine years and 75 percent are older than thirty-nine years.

In general, half the people have ages between the lower and upper quartiles, between thirty-nine and sixty-one. That is typical of the variation in age among the 1,947 people too: half are between thirty-nine and sixty-one, a quarter are below thirty-nine, and a quarter are above sixty-one. So we now have a typical age, fifty years, and a typical range of ages, thirty-nine to sixty-one.

The median and quartiles tell us more. For the 1,947 people, the lower quartile of age, thirty-nine years, is eleven years below the median, fifty years, and the upper quartile, sixty-one years, is eleven years above the median. Go up from the median eleven years and you capture 25 percent of the population, or go down from the median eleven years and you capture 25 percent of the population. The center of the distribution of ages looks symmetrical about the median. Symmetry is not a given. As we will see in a moment, income is not symmetrical: the lower quartile is closer to the median than the upper quartile. For those of us who are typical, near the median, the poor are much closer than the rich. So the median and quartiles tell us about the presence or absence of symmetry in the center of the distribution.

Typically, people vary. Look at a crowd of people at a stadium and you will see many typical variations. The quartiles of age, thirty-nine and sixty-one, tell us about the way the ages typically vary. Some people stand out in a crowd. They don't just vary; they are downright odd, perhaps in a good way, perhaps in a bad way, or perhaps in an interesting way, but in any event in a way that merits individual attention. The boxplot may call attention to a small number of individuals who stand out, who vary more than is typical. The boxplot creates an upper and lower boundary, and then it plots individuals as distinct points if they are outside the boundaries. Half the ages fall outside the quartiles, below thirty-nine or above sixty-one, so there is nothing odd about being outside the quartiles. What is the standard for being odd? As in life, so too in boxplots: the standard for being a little odd is somewhat arbitrary. Common outside boundaries are formed by calculating the difference between the quartiles, $61 - 39 = 22$, multiplying it by 1.5 to get $33 = 1.5 \times 22$, adding that to the upper quartile to get the upper boundary of $61 + 33 = 94$, subtracting from the lower boundary to get the lower quartile, $39 - 33 = 6$. Because NHANES capped age at eighty, and obtained periodontal measures only for those thirty years old or older, no one is outside the boundaries for age. No one is plotted as an individual.

We now have the ingredients of a boxplot: the median, upper and lower quartiles, and a definition of "outside" beyond which individuals are plotted separately. Figure 1 is a boxplot of the ages of the 1,947 people. We see everything at a glance. The horizontal center line is at the median, fifty: half the people are older and half are younger. Half

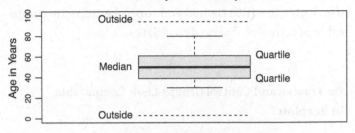

Example of a Boxplot

Ages of 1,947 People

Figure 1 An example of a boxplot, showing the median and upper and lower quartiles of age for 1,947 people. It happens that no points are plotted individually because no points are beyond the outside barriers.

the people are between the quartiles, inside the box, between thirty-nine and sixty-one. A quarter of the people are above the box, older than sixty-one. A quarter are below the box, younger than thirty-nine. A quarter of the people are inside the box and above the median of fifty, and a quarter are inside the box and below the median of fifty. There are no outside points. In this example, I have drawn lines at the boundaries that define outside points so that you can see them this first time, but these lines do not appear explicitly in a standard boxplot. Short lines, called whiskers, extend up and down from the box to the largest and smallest points that are not outside.

Boxplots let you see a batch of numbers at a glance and thus they are particularly helpful when you need to compare several batches of numbers. Boxplots can depict the degree to which treated and control groups are comparable in terms of observed covariates.

Do Treated and Control Groups Look Comparable in Boxplots?

Boxplots can depict the degree to which treated and control groups are comparable in terms of observed covariates. Figure 2 does this. For three covariates—age, income, and education—the figure compares 441 daily smokers (S) and 1,506 controls (C).

At a glance, figure 2 shows that the smokers were a few years younger than the controls, and that the smokers had much lower income and less education. The median smoker had a family income only slightly above the poverty level and had a high school degree, while the median control earned more than twice as much and had "some college," which might consist of an associate or technical degree from a community college.

Again, the situation is far worse than it appears in figure 2 because the figure considers one covariate at a time. What would the situation look like if we took all the covariates into account all at once? The covariates we have considered are age, sex, income, education, and race. Using all of this information, you could do a better job of

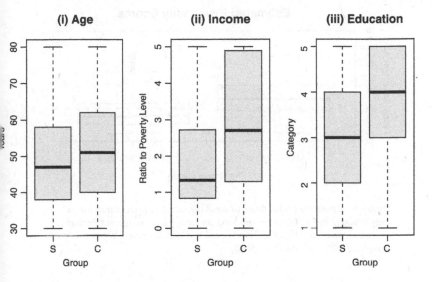

Figure 2 Comparison of daily smokers (S) and controls (C) in terms of age, income, and education. Income is the ratio of family income to the poverty level. Education is coded: 1 for < ninth grade, 2 for at least ninth grade without a high school degree, 3 for a high school degree or the equivalent, 4 for some college, and 5 for at least a four-year college degree.

guessing who will smoke than you could do using any one covariate. Figure 3 depicts the estimates, described earlier, of the probabilities of smoking for the 1,947 individuals. These estimates reflect the covariates age, sex, income, education, and race in a unified way. The probability of treatment—here, smoking—for a person with specified values of the observed covariates—here, age, sex, income, education, and race—is called the propensity score.

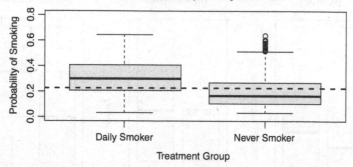

Figure 3 Estimated probabilities of smoking, or propensity scores, for 441 daily smokers and 1,506 never-smoking controls, based on age, sex, income, education, and race. The dashed horizontal line is at the proportion of smokers, $0.227 = 441/(441 + 1,506)$.

In figure 3, the smokers and controls are quite different. As expected, the estimated probabilities of smoking are higher for smokers, but the figure shows how much higher. The median probability in the control group, 0.166, is well below the lower quartile in the smoking group, 0.207, and this kind of separation did not occur in figure 2, where the covariates were considered one at a time.

As noted earlier, the largest estimated probability in figure 3 is more than twenty times the smallest probability. In contrast, had the treatments been assigned by the tickets-in-a-barrel randomization, the assignment probabilities would have all equaled 0.227, represented by the dashed horizontal line in figure 3.

The probability of treatment—here, smoking—for a person with specified values of the observed covariates—here, age, sex, income, education, and race— is called the propensity score.

The probabilities in figure 3 are estimates of the propensity score. The propensity score is the probability that a person with certain specified attributes will receive the treatment. The propensity score will be discussed in chapter 4.

Comparing Outcomes in Groups That Are Not Comparable

The outcome is a measure of periodontal disease. Excluding wisdom teeth, each of twenty-eight teeth was examined, if present, at each of six locations on each tooth, making 28 × 6 = 168 locations. The outcome for one person is the percent of locations exhibiting periodontal disease, 0 to 100 percent for each person. A location exhibits periodontal disease if there was either a pocket depth or loss of attachment of at least four millimeters, indicating a separation of the gums from the teeth.

Figure 4 compares periodontal disease for smokers and controls. At a glance, it is evident that the smokers have far more periodontal disease than the controls. For the nonsmokers, the median was 1.2 percent diseased locations, but for the smokers the median was more than ten times higher, or 12.4 percent diseased locations. At the upper quartiles, the difference is dramatic. A quarter of

Periodontal Disease

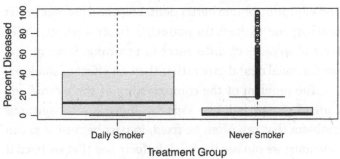

Figure 4 Periodontal disease in 441 daily smokers and 1,506 never-smoking controls.

the smokers had more than 42 percent diseased locations, whereas a quarter of controls had more than 7 percent diseased locations. Some of the controls had extensive periodontal disease, but they are few in number; they are outside points in the control boxplot.

The big question is, What does figure 4 tell us? Certainly, smokers have more extensive periodontal disease than nonsmokers, but why? Is this an effect caused by smoking? Or does the figure simply compare people who are not comparable? Is figure 4 a mere triviality, the mere finding that people who are not comparable are different? Figure 2 told us that the smoker boxplot in figure 4 has an excess of men with limited income and education, while the control boxplot has an excess of women with more income and education. Perhaps people with more income and education obtain better dental care and practice better oral hygiene. Perhaps the pattern in figure 4 reflects the effects of decades of differences in brushing, flossing, and professional dental care rather than an effect of smoking.

The problem of the comparability of the treated and control groups has two aspects. There is a conspicuous problem that can often be fixed, in part because it is conspicuous; we can see it, attack it, fix it, see that we fixed it. Alas, there is also an inconspicuous and at times invisible aspect that is harder to address.

The conspicuous aspect is evident in figures 2–3. The smokers and controls are visibly different in terms of the covariates that are measured. Chapter 4 will fix this conspicuous problem, comparing people who look comparable in terms of the covariates that were measured.

There are always covariates that were not measured. Always. Smokers and nonsmokers differ in terms of other addictive behaviors, consumption of alcohol and narcotics, and possibly personality and genetics, perhaps with widespread consequences for behavior and disease.[8] Chapter 5 discusses methods that address unmeasured covariates.

Problems from measured covariates are often fixed, and are often seen to be fixed. Problems from unmeasured covariates can be diminished or partially addressed, not eliminated, lingering on in diminished form, sometimes diminishing to the point of irrelevance. Because of this, the first task of control for measured covariates can seem to be only a minor part of causal inference in observational studies, with the overriding problem being the second task of addressing unmeasured covariates. If, however, we include in the first task the careful measurement of important covariates, then the two tasks are of comparable importance.

ADJUSTMENTS FOR MEASURED COVARIATES

Matching for Covariates as a Method of Adjustment

In figure 4, we saw more extensive periodontal disease among smokers, but we were not convinced that we were witnessing an effect caused by smoking. The figure compared the periodontal disease outcomes of treated individuals and controls who were not comparable. In figures 2–3, we saw that the smokers and nonsmokers were not comparable. The simplest solution is to compare individuals who are comparable, or at least comparable in ways we can see. True, people who look comparable in ways we can see might differ in ways we cannot see—ways that were not measured—but this chapter is concerned with the problem we can see.

In pair matching, the 441 smokers are each paired with a different control selected from the 1,506 controls.

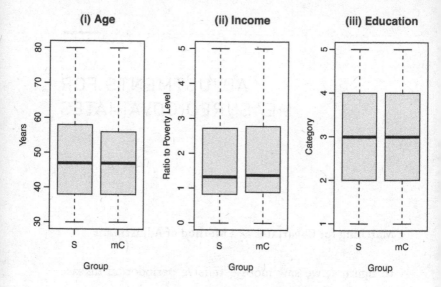

Figure 5 Three covariates for 441 smokers (S) and 441 matched controls (mC).

The goal is to select 441 controls who look similar to the smokers in terms of the measured covariates, sex, age, income, education, and race.

Figure 5 is analogous to figure 3, but it describes all 441 smokers and only their 441 matched controls. Unlike figure 3, the three covariates—age, income, and education—in figure 5 have similar distributions in treated and matched control groups. In both groups, the median age is forty-seven years. The median income is 1.33 times the poverty level for smokers and 1.38 times the poverty level for matched nonsmokers. The median education is a high school degree for both smokers and matched controls. The upper and lower quartiles are also similar in figure 5. For each of the three covariates in the figure, the entire distribution looks similar for smokers and matched controls.

There is less to plot with two-category covariates, but they too are balanced in the matched comparison. By design, each smoker was matched to one nonsmoker, so exactly 50 percent of the $882 = 2 \times 441$ matched individuals are smokers. The important question remains: Is this fifty-fifty split still present within subsets of people defined by the measured covariates? Is this true of men alone? Of women alone? Or of older women with a college education? Indeed, it is.

Gone is the pattern that the smokers were mostly men and the nonsmokers were mostly women: in the matched sample, 50.8 percent of the men and 49.9 percent of the women smoked. How does this imbalance in gender compare with that expected if $882 = 2 \times 441$ people are split at random into two groups of size 441? A larger difference in gender would occur with a high probability in a random split—specifically, with a probability of 0.63—so the balance for sex is comparable to that expected in a randomized experiment. The same is true for race: in the matched sample, 51.3 percent of the Black people were daily smokers, as were 49.4 percent of the rest, and again this imbalance is comparable to what is expected from random assignment. The same is true for older ages in the matched sample: 49.5 percent of those under sixty were smokers, as were 51.9 percent of those who were sixty or older.

There is also balance if we look at three covariates at once. Before matching, there had been a nearly tenfold difference in the proportion of smokers among women at least sixty with at least four years of college—only 4.3 percent were smokers—and among men under sixty without four years of college—fully 42.3 percent were smokers. After matching, that difference is gone: it is 50 percent smokers among women at least sixty with at least four years of college and 50.7 percent smokers among men under sixty without four years of college.

Imbalances in the Propensity Score

As we look inside smaller groups of people defined by more covariates, the data quickly become thin. Different people with different ages, incomes, educational backgrounds, genders, and races had different probabilities of becoming smokers, and we estimated these probabilities, or propensity scores, and plotted them in figure 3. In that figure, before matching, smokers and nonsmokers looked very different. How do they compare after matching?

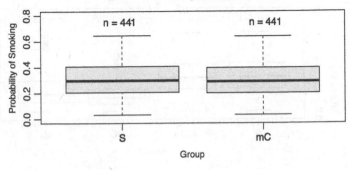

Figure 6 Propensity scores for 441 smokers (S) and 441 matched nonsmoking controls (mC).

Figure 6 compares the estimated probabilities of smoking after matching. Unlike figure 3, after matching, the boxplots of these propensity scores look similar.

Figure 7 looks more closely at the propensity score and thus helps in understanding how matching works. The first and last boxplots in figure 7 are the two boxplots in figure 3, the 441 smokers (S) and all 1,506 nonsmoking controls (aC). The first and second boxplots in figure 7 are the two boxplots in figure 6, the 441 smokers and 441 matched nonsmoking controls (mC). The new boxplot (uC) in figure 7 describes the 1,065 controls who were excluded from the matched sample, so 1,506 = 441 + 1,065. In words, the match found among the 1,506 nonsmokers a total of 441 nonsmokers who look like smokers, discarding the 1,065 nonsmokers who look nothing like smokers.

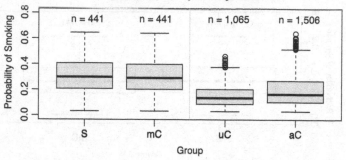

Figure 7 How matching changed the boxplot of propensity scores. The groups are S = smokers, mC = matched controls, uC = unmatched controls, and aC = all controls.

How Matching for the Propensity Score Balances Measured Covariates

There is a sense in which one measured covariate, the propensity score, summarizes all measured covariates, here age, sex, income, education, and race. There is a sense in which the balance in figure 6 for the propensity score signifies success for all measured covariates. And there is a sense in which matching for the propensity score fixes not only the problem in figure 3 but also those in figure 2. What is this sense? This section describes one of the several basic facts about propensity scores.[1]

Imagine a world slightly more complex than the PALM trial in chapter 2, but also a world much simpler than the observational study of periodontal disease. In this newly imagined world, some people are assigned to treatment or control by the flip of a fair coin, with heads for treatment and tails for control. Other people are assigned to treatment or control by the roll of a fair die, with a one sending them to treatment and two–six sending them to control. Let us call them coin people and die people, and to keep the arithmetic simple, let us suppose half the people are coin people and half are die people. The coin people do not have much in common: some are old men, and others are young women. The die people do not have much in common: some are old women, and others are young men. The question is whether we can forget about the details—age and sex, say—focus on coin people and die people, and let chance worry about the details. That worked in chapter 2 for the PALM trial—everyone was a coin person, and chance balanced the covariates. Now, however, there are two types of people, coin people and die people, with different probabilities of treatment, one-half and one-sixth.

In our imagined world, the chance of treatment is one-half for the coin people and one-sixth for the die people, and half the people are coin people and half are die people, so in aggregate the chance of treatment is $(1/2)(1/2) + (1/2)(1/6) = 1/4 + 1/12 = 4/12 = 1/3$, and the chance of control is $1 - 1/3 = 2/3$. Most treated people are coin people, however, and most of the controls are die people. There are too many coin people—old men and young women—in the treated group and too many die people—old women and young men—in the control group. We are hoping to fix the problem by matching.

Suppose that we match coin people to coin people and die people to die people, and let luck do the rest. Will that work? Intuition suggests it might work. Taken alone, the coin people form a randomized experiment. Taken alone, the die people form a randomized experiment, albeit with a one-sixth probability of treatment rather than a one-half probability. Each randomized experiment should balance its covariates separately. We got into trouble when we merged the coin people and die people, but matching is one way to keep them apart. So it might work. Does it?

Pairing coin people to coin people could pair two people who are different. Old men and young women are both coin people, so it might pair an old treated man to a young control woman, for example. On the other hand, it might pair a young treated woman to an old control man. So in light of this, I have a question for you. Suppose I give you two coin people, an old man and a young woman, and tell you that one is treated and the other is a control, but I do not tell you who is who. The question is, Can you guess who received the treatment? Well, obviously, you *can* guess—you could base your guess on anything you prefer, such as a new coin flip—but there is no reason the coin should get it right. The real question is, Can you guess in a way that will beat the coin? It does not sound easy to beat the coin. The old man had a one-half probability of being treated, but so did the young woman. From the information given, there is no reason to think the old man is treated, and no reason to think it is the young woman. In fact, you cannot beat the coin. If you pair two coin people, one treated and the other a control, then the two people are equally likely to be the treated person. The coin pairs are like a randomized experiment: if there are many such pairs, then old treated men paired with young control women tend to balance out with young treated women paired with old control men.

Although it may feel odd at first, the same thing happens with the die people. I give you two die people, an old woman and a young man. I tell you that one is treated and the other is the control, but I do not tell you who is who. I ask you to guess who is treated and do better than a coin flip in your guesses. At first, it mistakenly feels like you might beat the coin's guess because the probability of treatment is one-sixth for die people. Die people are likely to be controls. Can you use that fact? The dilemma you face is that both of them were die people, the old woman and the young man, so there is no reason to favor either of them in your guess. The situation with two die people feels different from that with two coin people because I told you more when I told you that two die people include one treated person and one control. If I flip a coin twice for two coin people, then half the time I get one treated person and one control, a quarter of the time I get two treated people, and a quarter of the time I get two controls, so with coin people, being told there is one treated person and one control is being told nothing surprising. If I roll a die twice to assign treatments to two die people, then with probability $(5/6)(5/6) = 25/36 = 0.694$ I get two controls. So when I told you a pair of die people includes one treated person and one control, I told you that pair was a bit unusual. In pair matching, as in figure 7, you insist that each pair contain one treated person and one control, and the question is what you will see in such a pair. If you form many pairs of two die people, one treated and the other a control, then pairs with an old treated woman paired to a young control man will occur, but they will happen with the same probability as a pair of a young treated man with an old control woman. With many die pairs and many coin pairs, age and sex will balance.

The situation is similar in figure 7. Suppose that, in figure 7, a treated person is paired with a control person who has the same propensity score—the same probability of being a smoker based on age, sex, income, education, and race. Those two people might be very different—that is, different in age, sex, income, and so on—but because they have the same propensity scores, those specific differences will not help you guess who is the smoker in the matched pair. If the propensity score is the same for both people in each matched pair, then the differences in age, sex, income, education, and race will tend to balance out when there are many matched pairs.

To illustrate this, consider two actual people in figure 6. Both people have propensity scores of 0.20 or a one-fifth chance of being a smoker based on age, sex, income, education, and race, yet just one person was in fact a smoker. One person is a woman aged forty-nine with a high school degree and a family income of 1.97 times the poverty level. The other is a man aged fifty-two with two years of college and a family income 4.07 times the poverty level. Both are not Black people. Men are more likely than women to smoke, but wealthier, better-educated people are less likely to smoke, and for these two people, these two conflicting tendencies balance one another perfectly to yield the same one-fifth chance of being a smoker. Given the information that one of these two people is a smoker, the detailed information about age, sex, income, education, and race is of no help in guessing which individual is the actual smoker. (It was the man.) All the information from age, sex, income, education, and race that might help to pick out the smoker is already in the propensity score of one-fifth that was computed from this information. Matching for one variable, the propensity score, balances the variables used in the construction of the propensity score. Propensity scores are often used to balance dozens or hundreds of covariates.

What does covariate balance look like? Figures 5–6 are one answer. In those figures, the smoking and matched control groups look comparable in aggregate for age, education, income, and the propensity score. Another answer slices figure 6 using the propensity score and looks at a covariate, say age. If we look at everyone in that figure with a propensity score in an interval of length 0.05 centered at $1/5 = 0.20$, then there are 48 smokers with a median age of 47.5 and 47 nonsmokers with a median age of 48. If the interval is again of length 0.05 but is centered at $1/10 = 0.10$, then there are 30 smokers with a median age of 52.5 and 28 nonsmokers with a median age of 51.5.

The first paragraph of this section said, "In a sense, in a sense, in a sense." In a sense, the propensity score computed from age, sex, income, education, and race tends to balance all five covariates. There are caveats. Obviously but most important, if the propensity score is computed from age, sex, income, education, and race, then it cannot be expected to balance some other covariate, perhaps a covariate you forgot to measure, such as the quality of the dental care that a person had as a child. Second, figure 7 depicts estimates of propensity scores, not true probabilities. Do a terrible job when estimating the propensity score, and the estimates may not do what the true probabilities would have done. Third, like randomization, the propensity score lets luck do most of the work when balancing the covariates, so it can balance a common attribute like sex if the sample size is not small, but it cannot balance attributes that are exceedingly rare or unique, like being Washington.

Comparing Outcomes While Controlling Measured Covariates

Figure 8 has the same format as figure 7, but attention shifts to the outcome—that is, the extent of periodontal disease. The outcome is the percent of tooth locations exhibiting periodontal disease, namely a separation of the teeth and gums. There are four boxplots in figure 8, one for 441 smokers (S), one for 441 matched controls (mC), one for 1,065 unmatched controls (uC), and one for all 1,506 controls (aC), where $441 + 1,065 = 1,506$.

The first and last boxplots in figure 8 for smokers (S) and all controls (aC) are the same as in figure 4, but we had trouble making sense of figure 4. In figure 4, the smokers and controls were quite different in terms of age, sex, education, income, and race, as seen in figure 2. We worried that people with more income and education might receive better professional dental care, practice better dental hygiene, or otherwise differ in terms of what else they smoked, drank, or ate. We worried that periodontal disease increases with age, and the smokers were younger. The smokers were disproportionately male. The boxplot for matched controls (mC) in figure 8 has removed the

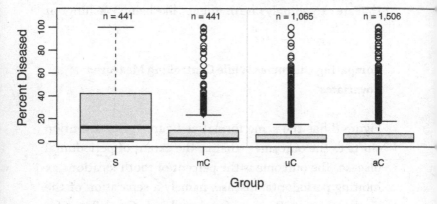

Figure 8 Extent of periodontal disease in smokers and matched controls. The groups are S = smokers, mC = matched controls, uC = unmatched controls, and aC = all controls.

differences in age, sex, education, income, and race, yet most smokers still have much more extensive periodontal disease than most nonsmokers. The extensive periodontal disease among smokers cannot be explained by differences in terms of age, sex, education, income, and race because nonsmokers who resemble smokers in terms of these covariates have much less extensive periodontal disease. Could it be something else, not an effect caused by smoking, but some other covariate? It could, but that is the topic of chapter 5.

If you squint at figure 8, you will see that the 441 matched controls (mC) have slightly more periodontal disease than the 1,065 unmatched controls. The difference is too large to be an accident of chance, but it is small in magnitude, on the order of half of 1 percent on the plotted scale of 0 to 100 percent. Matching for covariates did slightly alter periodontal disease among the controls, but the change is small compared with the level of periodontal disease among the smokers. In some other observational study, matching for observed covariates might eliminate what appeared to be an effect of the treatment.

Other Matching Techniques

A matched comparison may pair treated individuals to controls with similar propensity scores and do nothing more. There are reasons to do more, and more was done in the periodontal comparison. Matching for the propensity score tends to balance all measured covariates included in the score. The best matched samples result when propensity scores are combined with other matching methods.

As we have seen, two people may have similar propensity scores yet be quite different. In figure 7, there are twelve pairs in which the average propensity score, recorded to two decimals, is 0.16. The twenty-four people in these twelve pairs are very different, despite having similar propensity scores. Although none of the twenty-four was a Black person, there were twenty women and four men. Their ages ranged from thirty-four to sixty-six. Four people had high school degrees, but four others had at least a four-year college degree. Their incomes ranged from less than 2 times the poverty level to more than 5 times the poverty level. It seems wasteful and careless to regard these twenty-four people as the same—equally matchable—merely because their propensity scores are all close to 0.16. In fact, the match did not just pair people with similar propensity scores. It *always* paired people with similar propensity scores, but *whenever there was a choice* of several people with similar propensity scores, it picked people for the same pair to make them as similar as possible. For instance, one of the twelve pairs consisted of two women both forty-three years old, both with a high school degree, one with an income of 2.72 times the poverty level and the other with an income of 2.66 times the poverty level. Another pair consisted of two men with two years of college and incomes more than 5 times the poverty level, one aged fifty-three and the other aged fifty-four. And so on.

Propensity scores leave to chance most of the work in balancing covariates. Chance does a good job with big categories, such as men and women, or old and young. With obscure categories—say, women over sixty with some college and low propensity scores—individual categories may contain few people, so that chance imbalances are not under the control of the law of large numbers. Flip a

Matching for the propensity score tends to balance all measured covariates included in the score. The best matched samples result when propensity scores are combined with other matching methods.

coin 441 times, and the proportion of heads will be close to one-half, but flip it three times and there is a one-eighth chance of three heads and a one-eighth chance of three tails. The periodontal match forced the best-possible balance on a covariate with fifty-four categories built from sex, age above sixty, education, and categories of the propensity score. With 441 smokers and fifty-four categories, the average category count is only $441/54 = 8.2$ smokers, and fourteen categories had three or fewer smokers. You cannot rely on the law of large numbers if you flip a coin three times, so forcing balance in obscure categories is helpful.[2]

Matching is the simplest method of adjustment. With matching, the problem that treated individuals and controls differ visibly in measured covariates, as in figure 2, is fixed by comparing treated individuals and a subset of the controls who do not visibly differ, as in figure 5. Also, with matching, it is easy to get a sense of what is happening by looking at plots, as we have been doing. That said, there are many methods of adjustment, and it is not uncommon to use several methods at once.

SENSITIVITY TO
UNMEASURED COVARIATES

Some circumstantial evidence is very strong, as when you find a trout in the milk.

—Henry David Thoreau, journal, November 11, 1850

Objections, Counterclaims, and Rival Hypotheses

A symphony is followed by applause. An observational study is followed by objections. The common objection claims that the investigators adjusted for several covariates, but failed to measure and hence did not control another covariate. Had the investigators adjusted for this additional covariate, the objection continues, the ostensible effect of the treatment would have vanished. The common objection is sometimes reasonable, sometimes unreasonable, and often difficult to classify as reasonable or unreasonable. You can be misled by a mistaken claim derived from an observational study, but you can also be misled by a mistaken objection to such a study. Frequently, there is no way to err on the side of safety because either error has harmful consequences.

A first thought, understandable though incorrect, insists that rejecting scientific evidence as inadequate is a sign of discernment and high standards. This first thought views knowledge as an achievement, where higher standards for achievement are always better than lower ones. This first thought favors higher standards for knowledge, often leading to a denial that knowledge has been achieved. This first thought forgets that you can be accused of knowing, accused of having acted wrongfully given that you did, in fact, know.[1] For decades, the tobacco industry described as inadequate the evidence linking smoking with cancer and coronary disease; then in the 1990s, litigation forced the industry to pay enormous financial settlements in part because it knowingly raised doubts about what it knew to be true.[2]

How does one strike the right balance between scientific evidence and criticism of scientific evidence? In an extended essay that is worth reading in detail, Irwin Bross proposed the following answer:

In the great debate over smoking and lung cancer the quality of statistical criticism was, I think, rather poor (despite the eminence of the critics). . . . As a first step toward the ground rules of statistical criticism, let us examine the roles of the critic and the proponent. . . . Although the critic's role appears purely negative, it has a positive side to it. Implicitly (and sometimes explicitly) he puts forth a counterhypothesis. . . . [A] critic who objects to a bias in the design or a failure to control some established factor is, in fact, raising a counter-hypothesis (even though he may not state it). Since the counterhypothesis is essential in the logical structure of criticism, it facilitates debate when it is explicitly stated. . . . The critic has the responsibility for showing that his counterhypothesis is tenable. In so doing, he operates under the same ground rules as a proponent.[3]

Criticism of scientific evidence is part of science and must meet scientific standards. Criticism of scientific evidence is a counterhypothesis or counterclaim that explains the evidence in a different way, perhaps as biased treatment assignment rather than an effect caused by a treatment. Philosopher Ludwig Wittgenstein asked, "Doesn't one need grounds for doubt?"[4] In scientific work, grounds for doubt are part of the science. The grounds for doubt may be judged inadequate. A counterclaim may be rejected as vague, unsubstantiated, untestable, implausible, or motivated by self-interest, arrogance, or animosity. Bross concludes, "My theme has been: we should not have a 'double standard' in science and statistics, one standard for proponents and another for critics."[5]

Smoking and Lung Cancer

In the 1950s, the effect of cigarette smoking as a cause of lung cancer was controversial and widely debated. Observational studies had found a strong association between smoking and lung cancer.[6] Was this a biased association or were cigarettes a cause of lung cancer?

The inventor of randomized experimentation, Ronald Fisher, was a vocal critic of the early observational studies. In 1957, the *New York Times* reported,

> Sir Ronald A. Fisher, Arthur Balfour Professor of Genetics at Cambridge University, England, termed "inconclusive" the evidence turned up so far that would link cigarette smoking to lung cancer. . . . Sir Ronald has been credited with formulating many of the principles which today govern experimentation in the natural sciences and is noted for his theories of mathematical statistics and mathematical genetics. . . . "The evidence linking cigarette smoking with lung cancer, standing by itself," Sir Ronald

said, "is inconclusive, as it is apparently impossible to carry out properly controlled experiments with human material. Observations not fulfilling the requirements of decisive experimentation might be suggestive, not conclusive."[7]

Scientists have accepted Fisher's method—the random assignment of treatments—as the sound basis for causal inference. At the same time, they have tacitly rejected Fisher's standard, namely that if ethical or practical concerns bar randomized experimentation with humans, then inference about causal effects can be merely suggestive, never conclusive. Scientists tacitly rejected this standard when in the absence of randomized experiments on humans, they accepted as conclusive the evidence that smoking causes lung cancer. No one today regards smoking and lung cancer as an open question in need of further investigation due to some defect in the extant body of scientific evidence. That observational studies and other sources can sometimes provide adequate evidence of cause and effect does not mean this happens often, easily, quickly, or without extended controversy, but it does happen. Understanding how it can happen is the focus of the remainder of this book.

Observational studies routinely meet with objections, counterclaims, and rival explanations for the observed association between the treatment received and outcome exhibited. In this light, the existence of objections tells us little, so their content must be examined, perhaps over an extended period. In essayist Ralph Waldo Emerson's words, a person "must know how to estimate a sour face."[8] Sensitivity analysis is one tool for that task.

The First Sensitivity Analysis in an Observational Study

The first sensitivity analysis by Jerry Cornfield and colleagues occurred in a long discussion of the evidence available in 1959 implicating smoking as a cause of lung cancer. The evidence was varied. For instance, controlled experiments had produced skin cancer in mice using the tars in tobacco, and tobacco smoke had produced precancerous transformations in the lungs of mice and dogs. Smoking was associated with lung cancer in humans. Shifts in smoking behavior in human populations were followed, after a time, by corresponding shifts in the rate of lung cancer. Of course, there were no randomized trials of human subjects.

The sensitivity analysis concerned observational studies of humans, in particular the claim that the observed association between smoking and lung cancer might be spurious due to some failure to adjust for some unmeasured covariate associated with both smoking and lung cancer. Based on a little algebra in an appendix to their paper, Cornfield and colleagues concluded that the un-measured covariate would have to be quite remarkable to produce the observed association between smoking and lung cancer. They wrote,

> There is a quantitative question. Cigarette smokers have a nine-fold greater risk of developing lung cancer than nonsmokers, while over-two-pack-a-day smokers have at least a 60-fold greater risk. Any characteristic proposed as a measure of the postulated cause common to both smoking status and lung-cancer risk must therefore be at least nine-fold more prevalent among cigarette smokers than among nonsmokers and at least 60-fold more prevalent among two-pack-a-day smokers. No such characteristic has yet been produced despite diligent search.[9]

This calculation is an important conceptual advance. True, association does not imply causation: any observed association can be explained by a sufficiently large bias in treatment assignment due to a failure to control an un-observed covariate. To this, Cornfield and colleagues add a quantitative aspect: to explain the association actually seen in recorded data—to explain the association that needs explaining—the magnitude of the bias in treat-ment assignment needs to exceed a certain size. Joel

Association does not imply causation. Cornfield and colleagues add a quantitative aspect: to explain the association actually seen in data—to explain the association that needs explaining—the bias in treatment assignment needs to exceed a certain size.

Greenhouse put it this way, "No longer could one refute an observed causal association by simply asserting that some new factor (such as a genetic factor) might be the true cause. Now one had to argue that the relative prevalence of this potentially confounding factor was greater than the observed relative risk of the putative causal agent."[10] No longer could one say, "Anything can explain everything." Scientific counterclaims, like scientific claims, had to meet certain constraints imposed by empirical observations. In a tangible, quantitative sense, a sensitivity analysis achieved Bross's goal of a common standard for proponents and critics.

Although it is a key conceptual advance, this first method of sensitivity analysis is not adequate for general use. The method is limited to binary outcomes, so it does not apply to other common types of data. The method ignores the distinction between estimates from data and true population quantities, so it can produce misleading appraisals of observational studies of small or moderate sample size, where estimates may be unstable. Observational studies routinely adjust for measured covariates before discussing unmeasured covariates, but this first method presumes no such adjustments were made. This first method was tailored to the controversy about smoking and lung cancer, where the estimated effect was enormous, and in various senses, it tends to exaggerate sensitivity to bias in less dramatic contexts that are nonetheless important. Modern methods of sensitivity analysis fix these limitations. One of these methods is applied in the next section to the periodontal disease example from chapter 4.

Modern Sensitivity Analysis: Smoking and Periodontal Disease

One modern method of sensitivity analysis will be applied to the periodontal data from chapter 4.[11] Imagine a paired randomized experiment with 441 pairs, and a coin was flipped 441 times to pick one person in the pair for treatment, assigning the other to control. No matter what you knew about the two people in a pair before treatment assignment, each person would have a probability of one-half of being the treated person. Reasoning similar to that in chapter 2 would lead to tests of the hypothesis of no treatment effect, estimates of the magnitude of effect, and other common statistical inferences such as confidence intervals for the magnitude of effect. Of course, people made their own decisions about whether or not to smoke—they were not randomly assigned—so the reasoning in chapter 2 is not directly applicable. As was seen in figure 2, the people who chose to smoke were very different from those who chose to refrain from smoking: the smokers were younger, with less education and income, and were more often male. Presumably, the smokers differed in other ways as well. In chapter 4, matching removed the visible differences, but matching cannot be expected to remove differences that were not measured.

It is natural to think about an unmeasured covariate in much the same way that we think about a measured covariate. In chapter 4, the proportion of smokers was very different for women aged at least sixty with at least four years of college and for men under sixty without four years of college. In the first group, 4.3 percent were smokers, but in the second group 42.3 percent were smokers, or a tenfold difference. After matching, it was about 50 percent smokers in both groups. It is natural to think about an unobserved covariate in similar terms. In the matched sample, the probability of smoking might be higher for one person in a pair than for the other because they differ in terms of an unmeasured covariate.

Consider matched pair j, that is, the jth of the 441 pairs, perhaps pair $j = 247$ or some other pair. In a randomized paired experiment, the probability that the first person in pair j receives the treatment is $p_j = 1/2$, and the probability that the second person in the pair receives the treatment is $1 - p_j = 1/2$ because randomization assigns one person in the pair to treatment based on a flip of a fair coin that cares not at all about these two people and their attributes. It is natural to quantify the magnitude of the departure from a randomized experiment by the degree to which p_j and $1 - p_j$ may depart from a half. This probability p_j would depart from one-half if the two people in the pair had different chances of being the treated person in matched pair j because they differ in terms of attributes not controlled by the matching. Think about flipping 441 biased coins to assign treatments, where coin j comes up heads with a probability of p_j and tails with a probability of $1 - p_j$, where p_j need not be a half.

When we talk about coins and gambling, we speak of odds. If coin j comes up heads with a probability of p_j, then the odds of a head are $p_j/(1 - p_j)$. A fair coin is even money, one-to-one odds, or odds $(1/2)/(1 - 1/2) = 1/1$. If $p_j = 2/3$, then the coin is quite biased, and the odds of a head are two to one, or $(2/3)/(1 - 2/3) = 2/1$. If smoking did not cause periodontal disease, how biased would the coin have to be to produce the smoker (S) and matched control (mC) boxplots in figure 8?

The bias in treatment assignment is quantified by a number $\Gamma \geq 1$. It says that the bias in the coin is at most Γ, meaning that $p_j/(1 - p_j)$ is at most Γ, and $(1 - p_j)/p_j$ is at most Γ. If $\Gamma = 1$, then the coin is fair, with $p_j = 1/2$, as in a randomized experiment. If $\Gamma = 2$, then the coin could be quite biased, with p_j perhaps as large as two-thirds or perhaps as small as one-third. You could lose a lot of money quickly if you bet on a coin that you thought was fair—one-to-one odds or $\Gamma = 1$—when the true odds were two to one or $\Gamma = 2$.

The number Γ signifies a magnitude but not a direction of departure from one-to-one odds. The coins and probabilities p_j can vary from one pair to the next, but if $\Gamma = 2$, then the odds $p_j/(1 - p_j)$ are between one to two and two to one.

The sensitivity analysis asks a simple question, How large would Γ have to be for biased treatment assignment to produce the S and mC boxplots in figure 8? If in truth smoking had no effect on periodontal disease, what value of Γ could produce so many smokers with periodontal disease? Would a small bias—that is, a Γ close to one—explain away the ostensible effect of smoking in figure 8? Or would it take a very large bias, a Γ far from one?

If $\Gamma = 1$—that is, if figure 8 arose in a paired randomized experiment—then it would take a miracle to produce this figure if smoking did not cause periodontal disease. Figure 8 is logically possible in a randomized experiment with no treatment effect, but it is exceedingly improbable; indeed, its probability is so small that the computer cannot distinguish that probability from zero. Of course, figure 8 is not from a randomized experiment, so there is no basis for believing that $\Gamma = 1$.

Consider a bias in treatment assignment of $\Gamma = 2$, meaning that $1/3 \leq p_j \leq 2/3$ for every pair j. If each p_j were one-third or two-thirds, then this would be a substantial departure from random or equitable treatment assignment. If there were no effect of smoking, but each p_j equals one-third or two-thirds, then you could pick the winner two-thirds of the time; that is, in two-thirds of pairs, you could assign to smoking the person who will develop greater periodontal disease, thereby creating a false impression that smoking causes periodontal disease. And yet a bias of $\Gamma = 2$ is far too small to produce figure 8. If $\Gamma = 2$, then the probability of such a large ostensible effect of smoking in figure 8 is at most 0.000018. Either $\Gamma > 2$ or smoking really does cause periodontal disease.

Figure 9 is a picture of the situation described in words in the previous paragraph. The boxplot labeled P in figure 9 is the actual matched periodontal data from figure 8, now expressed as the 441 smoker minus control pair differences in periodontal disease, extending from –100 to 100 percent. In the boxplot labeled P, the differences are more often positive than negative, so the typical smoker has more extensive periodontal disease than their matched control. The three boxplots labeled S1, S2, and S3 are simulated data sets formed by flipping a biased coin with bias $\Gamma = 2$ in an imagined world with no treatment effect. The three S boxplots are built from the actual data, the hypothesis that smoking has no effect, and three sequences of 441 flips of a biased coin with $p_j = 1/3$ or $p_j = 2/3$. It is clear that the shift toward positive pair differences— toward greater periodontal disease in smokers—is larger in the actual data than in the three simulated data sets.

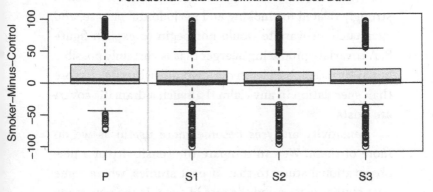

Figure 9 Comparison of the actual smoker minus control pair differences in periodontal disease (P) with three simulated samples (S1–S3) with no treatment effect and a bias of $\Gamma = 2$.

Figure 9 suggests that in the absence of an effect of smoking, a bias of $\Gamma = 2$ is just not big enough to produce what we saw in the actual data. A larger bias would be needed to produce the boxplot labeled P.

We have found that a bias of $\Gamma = 2$ is not big enough to produce the pattern of periodontal disease seen in figure 8. What does that mean? The numerical value of Γ can be understood in another way. Precisely the same bias can be described in terms of how an unobserved covariate relates to smoking and periodontal disease separately. A bias of $\Gamma = 2$ is produced by an unmeasured covariate that increases the odds of smoking by threefold and increases the odds of greater periodontal disease by fivefold. So a bias of $\Gamma = 2$ corresponds with an unmeasured covariate that is strongly related to smoking and periodontal disease, and even such a covariate would not begin to explain figure 8. A covariate producing a larger bias is certainly possible, but as Bross emphasized, a critic would owe us much more than speculation in any claim that such a dramatic covariate exists.

Sensitivity analyses become more useful as we do more of them. We can compare the sensitivity of a new observational study to that of past studies, where some past studies have stood the test of time. Is the new study as insensitive to bias as successful past studies? Smoking as a cause of lung cancer is insensitive to enormous biases and has stood the test of time; it is insensitive to $\Gamma = 5$. The use of a seat belt to prevent death in a car crash is also insensitive to enormous biases and has stood the test of time; it too is insensitive to $\Gamma = 5$.[12] A scientific article may make a splash in the newspapers, yet be sensitive to trivial biases, $\Gamma = 1.05$. Ruoqi Yu and colleagues discuss an observational study that was sensitive to small biases and was contradicted by subsequent randomized trials.[13]

The Role of Sensitivity Analysis

As noted when this chapter began, an observational study is met with objections, not applause. Empirical science is

A sensitivity analysis provides quantitative clarification, in light of empirical data, of what is being asserted by a proponent of a claim and by a critic of that claim.

an argument about what is true, conducted in the presence of experiments, observations, and data. Seeking counsel, we turn to Emerson again, who wrote, "Our culture, therefore, must not omit the arming of the man."[14] Sensitivity analysis can be a shield against frivolous counterclaims. Or it can reveal a claim to be fragile, highly dependent on the dubious assumption that people pick the treatments they impose on themselves by flipping a fair coin. A sensitivity analysis does not provide new empirical data; rather, it supplies quantitative clarification, in light of empirical data, of what is being asserted by a proponent of a claim and by a critic of that claim.

QUASI-EXPERIMENTAL DEVICES IN THE DESIGN OF OBSERVATIONAL STUDIES

Doesn't one need grounds for doubt? . . . One doubts on specific grounds.

—Ludwig Wittgenstein, *On Certainty*

Anticipated Counterclaims

Unlike a sensitivity analysis, a quasi-experimental device furnishes new data intended to advance a claim by undermining a specific counterclaim, by undermining specific grounds for doubt. Part of what makes a counterclaim credible is that the critic is raising a familiar problem—one that has led to mistaken conclusions in the past. Although a critic of an observational study may raise a surprising counterclaim, many, if not most, credible counterclaims can be anticipated before the study is begun, so the study may be designed to resist some anticipated counterclaims. Quasi-experimental devices are tactics intended to investigate and perhaps invalidate anticipated counterclaims. The systematic study of quasi-experimental devices started in 1957 with the work of Donald T. Campbell. What are examples of anticipated counterclaims?

Drugs, medical procedures, psychological counseling, economic assistance, and punishments for crimes may have unintended side effects, but problems and hence counterclaims are anticipated when side effects are studied outside randomized trials. If a person is given a drug, then a doctor likely thought the drug would be beneficial based on this person's symptoms. A control who did not receive the treatment likely had no symptoms, or different or less severe symptoms. This ambiguity is often called "confounding by indication," meaning that it is difficult to distinguish the effects caused by the treatment from the indications that the treatment was needed. If two people are convicted of spousal assault but only one is imprisoned, then there may be good reasons why they received different punishments. The very fact that one person was treated and the other was not might be taken as evidence that they were not comparable, even if they appear comparable in the available data. What additional data might address this counterclaim?

Unlike a sensitivity analysis, a quasi-experimental device furnishes new data intended to advance a claim by undermining a specific counterclaim, by undermining specific grounds for doubt.

A new public policy—a change in the tax code, increase in the minimum wage, or law placing restrictions on handgun purchases—often begins abruptly on a specific date stated in an act of legislation. For people who come under the scope of the policy, before that date, everyone is a control; after that date, everyone is a treated individual. A comparison of treated individuals this year with controls from last year faces the anticipated counterclaim that this year and last year differ in many ways; for instance, the weather was awful last year and people stayed home, but the stock market crashed this year and people felt poorer. The anticipated counterclaim is that the change in policy this year was only one difference between last year and this year, and perhaps a change in outcomes from last year to this year is not an effect caused by the change in policy. What additional data might address this counterclaim?

Two Control Groups

Azithromycin is an antibiotic drug. Other antibiotic drugs that are close relatives of azithromycin are thought to be associated with rare sudden cardiac death, possibly the result of arrhythmias. Using data from the Tennessee Medicaid program, Wayne Ray and colleagues asked whether there is an increase in cardiac death in the five days following the start of treatment with azithromycin. What is the natural control group? To whom should patients treated with azithromycin be compared?[1]

Ray and colleagues used two control groups. They compared patients receiving azithromycin to people who had received no antibiotic—that was the first control group. The second control group consisted of patients who had received another antibiotic, amoxicillin, that might be prescribed instead of azithromycin. An objection or counterclaim can be raised to either control group used alone.

Antibiotics are typically given to patients who are thought to have a bacterial infection, so most patients receiving azithromycin are likely to have an infection, whereas most people in the first control group have no symptoms of infection. Using the first control group alone, it would be difficult to distinguish the following two situations: azithromycin causes some cardiac deaths, or some infections cause some cardiac deaths. Excess cardiac deaths in the azithromycin group when compared to the first control group may not indicate that azithromycin is the cause of these deaths.

The second control group received an antibiotic, albeit a different one, so the patients in both the azithromycin and amoxicillin groups are likely to have a bacterial infection. This second control group removes a key problem with the first control group, but the amoxicillin control group has a problem of its own. If azithromycin and amoxicillin both caused cardiac deaths, and did so to the same degree, a comparison of these two groups could show no difference in cardiac mortality, even though azithromycin is causing deaths.

Together, the two control groups create a study design that is less ambiguous than either control group would be on its own. If there is an excess of cardiac deaths in the azithromycin group when compared to both control groups, then that finding is not easily dismissed as possibly caused by the infection rather than by azithromycin. After all, patients in the amoxicillin group also had an infection. If the azithromycin and amoxicillin groups are similar in terms of cardiac deaths, which occur at a higher rate than in the first control group, then we should exercise caution before attributing those deaths to treatment with azithromycin. Such a finding remains ambiguous: perhaps azithromycin and amoxicillin are both causing cardiac deaths, or perhaps the underlying infection is the cause, but there would be no reason to focus on azithromycin as the one antibiotic to avoid.

In fact, after adjusting for measured covariates as in chapter 4, Ray and colleagues found an excess of cardiac deaths in the azithromycin group when compared to each control group, both the untreated and amoxicillin controls.

The example just described is typical of a successful use of a quasi-experimental device. The most plausible counterclaim is anticipated and addressed by an additional component of data and an additional comparison.

The additional comparison undermines this most plausible counterclaim, but it does not eliminate all conceivable counterclaims.

The Logic of Two Control Groups

What attributes should we seek when we consider adding a second control group? To be of service, a second control must differ from the first in some pertinent way. I will mention one line of reasoning, due originally to experimental psychologist M. E. Bitterman and developed by social scientist Donald T. Campbell.[2] It is "control by systematic variation."

There is a covariate that is likely to be a basis for objections and counterclaims if nothing is done to preempt them. Perhaps the covariate is not measured or is inadequately measured. Instead of measuring the covariate and adjusting for it, the investigator systematically varies the covariate. That is, the investigator finds two control groups that clearly are very different with respect to this covariate, even though the covariate itself is not measured. In the study by Ray and colleagues, the covariate was the presence and symptoms of the infection, and it was clear that the untreated control group had fewer infections and fewer symptoms of infection than did the amoxicillin control group. Suppose the investigator finds that two control groups have similar outcomes despite differing greatly in terms of an unmeasured covariate, as happened in the study by Ray and colleagues. Then that finding tends to undermine a counterclaim saying the treated group and two control groups have different outcomes because of a failure to control for that covariate.

Untreated Counterparts in Addition to Untreated Controls

Working is expensive. There are transportation costs and often costs of childcare. A single mother with children may find that the cost of childcare exceeds what she would earn initially if she went to work. The Earned Income Tax Credit (EITC) is a form of negative income tax that subsidizes work by some individuals whose income is low. The EITC frequently elicits bipartisan support because it is designed to help the poor while encouraging them to work, with the ultimate goal of self-sufficiency. The Tax Reform Act of 1986 expanded the EITC starting in 1987. Did it work? What was the effect of this expansion on workforce participation? Did the expansion bring people into the workforce?

Nada Eissa and Jeffrey Liebman tried to answer this question using the US Current Population Surveys for 1985–1987, before the expansion of the EITC, and 1989–1991,

Suppose two control groups have similar outcomes despite differing greatly in terms of an unmeasured covariate. That tends to undermine a counter-claim saying the treated group and two control groups have different outcomes because of a failure to control for that covariate.

when the expansion was in effect. In one comparison, they examined unmarried women without a high school degree who had children. Many such women would have been eligible for increased benefits under the Tax Reform Act of 1986. In this group, workforce participation rose by 1.8 percent, from 47.9 percent in 1985–1987 to 49.7 percent in 1989–1991. Eissa and Liebman present both simple comparisons like this and ones that adjust for covariates, as in chapter 4; in the following brief discussion, however, I will describe some of the simple comparisons.

Was this 1.8 percentage point increase in workforce participation caused by the Tax Reform Act of 1986? It might have been, but many aspects of the economy move about from year to year. This treated versus control comparison is open to a counterclaim: perhaps the increase in workforce participation reflected some general economic trend, not any effect caused the Tax Reform Act of 1986. Of course, Eissa and Liebman anticipated this counterclaim, and tried to address it.

Most women were not eligible for the EITC. Women without children and those whose incomes were not low were typically ineligible for the EITC. Ineligible women were not directly affected by the EITC provisions of the Tax Reform Act of 1986, before or after 1987, so they provide some indication of general economic trends, unmixed with effects of the EITC. The ineligible women were, both by definition and in their behavior, different from the eligible women, so they are nothing like controls. In particular, these women were much more likely to be working, before and after 1987. I will call such women "counterparts" to distinguish them from comparable controls.

Eissa and Liebman considered two sets of counterparts. The first counterparts were unmarried women with less than a high school education but without children. The second counterparts were unmarried women with at least a high school degree with children. Some women in the second group would be eligible for the EITC, but far fewer such women would meet the income requirements for it. We look at the counterparts in an effort to see if and how they may have been affected by general economic trends.

The first counterparts experienced a decline in workforce participation of −2.3 percent, from 78.4 percent in 1985–1987 to 76.1 percent in 1989–1991. The second counterparts experienced an increase in workforce participation of 0.9 percent from 91.1 percent in 1985–1987 to 92 percent in 1989–1991. Again, some low-income counterparts in the second group may have been affected by the EITC.

The counterparts address concerns about general economic trends that affect everyone in a similar way. The increase of 1.8 percent in workforce participation in the 1989–1991 treated group, when compared to the 1985–1987 control group, is larger than both the −2.3 percent decline for the first counterparts and 0.9 percent increase

A quasi-experimental device strengthens a causal claim using data that weakens some anticipated counterclaim.

for the second counterparts. The 1.8 percent is not so easily dismissed as a general economic trend because that trend was not evident in the counterparts. A quasi-experimental device—two added comparisons involving counterparts—has weakened one natural counterclaim and thereby strengthened a causal claim.

Resolving Anticipated Counterclaims

A quasi-experimental device strengthens a causal claim using data that weakens some anticipated counterclaim. Quasi-experimental devices are the persistent diligence of the laboratory. Elements are added to the design of an observational study to resolve anticipated counterclaims. The correct explanation is found by eliminating mistaken explanations, one at a time.

NATURAL EXPERIMENTS, DISCONTINUITIES, AND INSTRUMENTS

A nostalgia for caprice and chaos.

—E. M. Cioran, *History and Utopia*

Bits and Pieces of Random Assignment in an Otherwise Biased World

The world contains bits of true randomness. There are lotteries that are truly random, in the same sense that a randomized trial is truly random. Some states in the United States and some nations in Europe play the role of the casino, running lotteries to raise money to support public services. A form of public assistance—such as subsidized housing—may be oversubscribed, so a lottery decides who receives subsidized housing and who does not. The same may be true of limited spaces for students in charter schools. These and similar situations are often called natural experiments. Is natural randomness useful? It sounds useful.

Two lotteries for spaces in charter schools, one in New York City and the other in Louisiana, give opposing conclusions about the effectiveness of charter schools.[1] Perhaps charter schools are not all the same, with some being more effective and others less so.

Nature has its own lotteries. Your mother has two not quite identical copies of each gene, and she gave each of her children one copy picked at random. That is, each of her egg cells contained one copy or the other of each gene, but which copy the eggs contained had nothing to do with which egg cell met a sperm cell from your father to make you. It is just luck that you and your brother have the same copy of a particular gene, but your sister has the alternative copy. The situation is almost, but not quite, the same for the genes received from your father. So you and your siblings constitute a small randomized experiment exploring the effects caused by the difference between your mother's two copies of each gene. Is that useful? It must be more complicated than that. Well, yes, actually, it is more complicated than that, but it is useful despite the complications.

Sometimes a lottery randomizes something, but what it randomizes is not the right something. In a study by Brian Jacob and Jens Ludwig based on a lottery for

The world contains bits of true randomness. Is natural randomness useful? It sounds useful.

subsidized housing, offers of housing subsidies were randomized by a lottery, but when the offers were made, many people turned them down.[2] It is straightforward to estimate the effects of the offer of a subsidy, because the offer was randomized. For most purposes, it is more interesting to know the effect of receiving a subsidy, but that was not randomized. How can a lottery be used if it randomizes something, but not the right something? The answer involves what are known as instruments or instrumental variables.

Natural Experiments from Lotteries

Does receiving a pile of cash reduce the risk of personal bankruptcy? Or is one dollop of cash beside the point for someone who has difficulty managing money? Scott Hankins, Mark Hoekstra, and Paige Marta Skiba addressed this question using data from the Florida Fantasy 5 lottery—a lottery run by the US state of Florida.[3] The Fantasy 5 lottery gave large prizes to individuals who correctly guessed five out of five random numbers, but under some circumstances, it gave much smaller prizes to individuals who correctly guessed four of the five numbers. Hankins and colleagues compared the rates of bankruptcy among the winners of large amounts, say $50,000 to $150,000, to the winners of much smaller amounts, say under $10,000.

They discuss 14,668 winners of under $10,000 and 1,212 winners of $50,000 to $150,000. In years zero through two following the win, the winners of large amounts had lower rates of bankruptcy than the winners of small amounts, but in years three through five, the large winners had higher rates of bankruptcy, so that in aggregate over years zero through five, the rates of bankruptcy were about the same.

Do housing subsidies from the US government encourage or discourage work? A first thought is that assistance can only help. A second thought is that perhaps the subsidy creates a disincentive to work because the subsidy is reduced as income from work increases. Brian Jacob and Jens Ludwig wrote, "Economic theory yields ambiguous predictions about the sign, let alone the magnitude, of any labor supply response to means-tested housing programs."[4] They found a natural experiment in Chicago, where in 1997 the Chicago Housing Authority Corporation had too few housing vouchers to meet the demand, so a collection of 82,607 eligible applicants was randomized to positions on a waiting list, with vouchers offered first to those at the top of the waiting list. By 2003, vouchers had been offered to 18,100 families on the list. One aspect of their analysis compared families offered a voucher and those not offered a voucher. The offer of a housing voucher was determined by position on the waiting list, and that had been randomized. Jacob and Ludwig found there was a small decline in employment and earnings among heads of households who were offered a voucher, when compared to people further down the waiting list who were not offered a voucher. Later, we will consider the effect of accepting the offer of a voucher—something that cannot be randomized.

Nature's Natural Experiment, I: The Genes of Siblings

One of the many molecules that your body manufactures is cytotoxic T lymphocyte antigen-4 or CTLA-4, and it plays a role in regulating the activity of your immune system. In your DNA is a gene, also called CTLA-4, that describes how to make the molecule CTLA-4. The gene is a set of instructions for manufacturing an important molecule. The gene is a long sequence of molecular letters in an alphabet with four letters. Your mother has two copies of this gene, and so does your father. As it turns out, there are two versions of this gene—let us call these versions A and a—that differ in a letter. Imagine two pages of text, text A and text a, that are almost identical except in one particular spot: one letter in text A has been changed to another letter to yield text a. Imagine that text A was written in English in the United Kingdom and text a was written in the United States, so the two texts are almost identical except for one letter in one word that is spelled differently in the

United Kingdom and United States. Sometimes changing one letter in a long set of instructions does not change the molecule that is built from those instructions, and sometimes it changes the molecule but in an innocuous way. Yet sometimes changing one letter profoundly changes how the molecule functions. Bijayeswar Vaidya and colleagues asked whether the difference between A and a plays a role in causing Graves' disease, an autoimmune disease involving the thyroid.[5]

Vaidya and colleagues looked at pairs of siblings, one of whom had Graves' disease. Figure 10 is an imagined example of one pair of siblings, where the mother has one copy of A and one of a, and the father has two copies of A. In this particular sibship, each child is sure to receive A from the father, and receives A or a, each with a probability of one-half, from the mother. By luck in this sibship, one daughter turned out to be AA and the other to be Aa. Regardless of the parents' genes, the genetic makeup of the two daughters might just as easily have been reversed: in this case, daughter 1 might have been Aa and daughter 2 might have been AA. It is in this sense that the genotypes actually present in two siblings resemble the random assignment of those two genotypes to those two children. This possible random reversal of genotypes—this so-called exchangeability—within a pair of siblings is the basis for several tests developed by David Curtis and by Richard Spielman and Warren Ewens and others

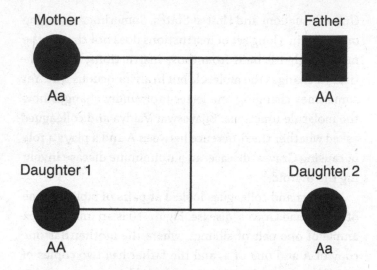

Figure 10 One sibship composed of two parents and two daughters. The mother has versions A and a of CTLA-4, while the father has two copies of version A. Both daughters receive A from the father, and they have a probability of one-half of receiving A or a from the mother.

of the hypothesis that a genetic marker—a one-letter difference—is closely linked to a genetic cause of a disease.[6] If we saw, as Vaidya and colleagues did, that the sibling with Graves' disease typically had an excess of A genes and those without Graves' disease typically had an excess of a genes, then we would have good reason to believe that something near the A/a marker on the CTLA-4 gene plays some role in causing Graves' disease.

There are several types of randomness in figure 10, and the comparison of siblings has used one type. The first daughter in the figure was AA while the second was Aa; with the same probability, however, the reverse situation could have occurred, namely the first daughter is Aa and the second is AA. The probabilities are said to be exchangeable: exchange the daughters and the probability does not change. That is, for CTLA-4 and the entire genome, this mother and father might have produced daughter two first and then daughter one with the same probability—whatever it was—that they produced daughter one followed by daughter two. We are speaking here about their genes; obviously, being born into a family with an older sister may affect what happens after birth.

We know the probabilities for the daughters in figure 10 are exchangeable without looking at the genes of the parents. This is important in studies of diseases that typically occur late in life, such as Alzheimer's. By the time a daughter develops Alzheimer's disease, the parents may be long dead, so the genes of the parents may be unavailable when the study is done. It is therefore convenient that the daughters are exchangeable regardless of the genomes of their parents; the parents are not needed when comparing the siblings.

For example, Stavra Romas and colleagues compared cases of Alzheimer's disease to sibling controls in terms of the frequency of the APOEε4 variant or allele of the APOE gene.[7] If this allele were unrelated to Alzheimer's disease, an equitable sharing of the APOEε4 alleles among siblings leads to an expectation of 57.78 APOEε4 alleles for the Alzheimer's cases, whereas 74 were observed. If the APOEε4 allele were irrelevant to Alzheimer's, such a large excess of the allele among cases, when compared to siblings, would occur only with probability 0.00000579.

So in genetics, the comparison of siblings forms a natural experiment, even if the genetic information from the parents is not available. What if the parents are available, but there are no siblings?

Nature's Natural Experiment, II: Hypothetical Siblings

The comparison of daughters one and two used one bit of randomness in figure 10, but there are other bits that are equally useful. If we knew the parents' genes in that figure, then we would know that every child of these parents will be either Aa or AA, each with a probability of one-half. We would know this even if both daughters had actually been AA, as they might have been. We would know this even if there was only one daughter.

The situation is only slightly more complex if both parents had been Aa. In this case, the mother and father can both contribute A, yielding a child who is AA. Or the mother and father can both contribute a, yielding a child who is aa. Or the mother can contribute A and the father can contribute a, yielding a child who is Aa. Or the mother can contribute a and the father can contribute A, again yielding a child who is Aa. What determines what happens? It depends on which sperm cell meets which egg cell, and that is essentially random. The four cases have an equal probability of one-fourth, so AA occurs with a probability of one-fourth, aa occurs with a probability one-fourth, and Aa occurs with a probability of $1/4 + 1/4 = 1/2$. We know this even if there is only one child who, by luck, happens to be AA.

Suppose that we had genetic information for triads—two parents and a child—where the child has some disease diagnosed early in life, such as autism. We need an early diagnosis because we need genetic information from the parents. Someday genetic information may be stored electronically forever for everyone, and then an early diagnosis would not be needed.

Given genetic information for a triad—two parents and a child with a disease—we have what is needed to test that a genetic variant or allele, A or a, is close on the chromosome to a genetic cause of the disease. This method is due to Richard Spielman, Ralph McGinnis, and Warren Ewens, and it is called the transmission disequilibrium test (TDT).[8] The test says that if the allele is not close to a genetic cause of the disease, then children with the disease should be AA, Aa, or aa with the probabilities determined by their parents, as illustrated in figure 10. If those probabilities do not describe the diseased children—if there is an excess of A's and deficit of a's, for example—then that is evidence of a genetic cause near A/a on the chromosome. How else can an excess of A's at conception be associated with later disease?

The remarkable thing about the TDT is that every child under study has the disease. Diseased children are not compared to other children who are free of the disease. The diseased children are compared to their hypothetical siblings—all the siblings these two parents could produce. Two actual siblings, as in figure 10, reflect some luck: instead of figure 10, both daughters could have been AA or Aa. In contrast, the possible children that these two parents can produce, and the relative frequencies of these children, are known once the parents' genotypes are known.

Joseph Dougherty and colleagues used the TDT to examine two variants of the Celf6 gene and its possible role in autism.[9] One variant was present at a higher frequency in autistic males than you would expect based on their parents' genomes. This is a complex study, not least because quite a few genes were examined.

Discontinuity Designs as Natural Experiments

Lotteries are sometimes used to equitably distribute a scarce resource, but they are not common. More often, it is first come, first served. Or it is those who need it most are most likely to get it. Or something else puts people in order, with those ordered first receiving the treatment and those ordered last receiving the control. Is there anything like a natural experiment in such a situation?

You want to buy tickets for a concert. You suspect it will sell out, so you arrive three hours early. Already there is a big line. You can see that the people at the head of the line have sleeping bags; they camped out to be sure they got tickets. After two hours of waiting, you look behind you and see that the line has doubled in length, and people keep joining it. The people at the front of the line are very different from those who just joined the line; camping out for tickets is different from whimsically seeing if there are any tickets left. The line starts to move, tickets are sold, and the people at the front are getting their tickets. You hope that you will reach the ticket office before they run out of tickets. At some point—before or after you reach the ticket office—someone comes out of the ticket office and waves people away, saying, "No more tickets, the ticket office is closed." This does not sound like a random assignment of tickets. Does any aspect of this situation resemble random assignment?

At some moment, one couple bought the last two tickets and the next couple watched the door to the ticket office close in front of them. Both couples hoped for tickets and worried they might be locked out, but as luck would have it, the first couple got tickets and the second did not. The good luck of the first couple and bad luck of the second is not quite random assignment—the first couple did join the line a moment earlier—but it is close to random assignment. Neither couple camped out, but neither couple arrived whimsically late. A comparison of people who received tickets and those turned away would compare incomparable groups, those with sleeping bags and those that joined the end of the line as the ticket office closed. A comparison of the last people to get tickets and the first people to be shut out—a comparison of people on opposite sides of the door as it slammed shut—is a much more equitable one.

The situation just described is the basis for the "discontinuity design" of Donald Thistlethwaite and Donald Campbell.[10] Assignment to treatment ends and assignment to control begins abruptly, at a discontinuity, along some continuum, such as when the door slammed shut in the concert example. Near the discontinuity, there is a natural experiment; far from the discontinuity, it is nothing like a natural experiment. The idea is that people are systematically different as you move along the continuum, but the systematic changes occur gradually, while the switch from treatment to control is total and abrupt; the door slams shut.

John DiNardo and David Lee used a discontinuity design to study the effect of unionization on wages. Do unions obtain higher wages for workers? Do they impose higher costs on employers? Unionized firms differ from nonunionized ones in many ways. Is there a natural experiment? DiNardo and Lee observed, "Our analysis is based on the fact that most new unionization occurs as a result of a secret ballot election. . . . This process creates a natural set of comparisons between establishments that faced elections where the union barely won . . . and those that faced elections where the union barely lost."[11] They found little evidence that labor organizing drives in 1984–1999 caused increased wages for union members, but also little evidence that they imposed higher costs on newly unionized establishments.

Discontinuities can be lines on a map rather than a point on a line. Children who live on opposite sides of the same street may be forced to attend different public schools, one better than the other. What do parents think the better public schools are worth? Sandra Black compared home prices on opposite sides of streets that define the limits of school zones.[12]

In some states and cities of the United States, voters can place legislative statutes on the ballot for passage by voters. Instead of just voting for a legislator in the hope that the legislator will enact a law, voters put the law on the ballot and enact it themselves. Do such ballot initiatives cause higher turnouts in elections? Luke Keele, Rocio Titiunik, and José Zubizarreta examined this question by comparing voting near the boundaries of adjacent voting districts with and without ballot initiatives.[13]

Encouragement Experiments: Can You Learn about One Treatment When You Randomize Another?

The simplest example of an instrument occurs in Paul Holland's randomized encouragement experiment, such as encouragement to quit smoking.[14] Perhaps encouragement is compared with the absence of encouragement, or encouragement to do one thing is compared with encouragement to do something else, or perhaps different forms of encouragement to do the same thing are compared. By definition, a randomized encouragement experiment has three distinctive features: the investigator randomly assigns types of encouragement to individuals in the experiment; the investigator does not control whether individuals heed encouragement and do what they are encouraged to do; and encouragement has no effect on those individuals who ignore it and do what they would have done anyway. The third condition has a technical name: the "exclusion restriction." It may or may not be true, and it is only relevant to one of the several questions an encouragement experiment may set out to answer. If we were studying the effect of encouragement to quit smoking on a biological measure of lung function, then it seems plausible that encouragement to quit smoking affects lung function only to the extent that encouragement causes a change in smoking behavior, and so the exclusion restriction seems plausible. In contrast, if we were studying the effect on some measure of contentment, then it seems possible that the third condition is false: being encouraged to quit smoking and failing to do so may be frustrating and disappointing, so it might affect contentment.

There are two possible questions in an encouragement experiment. What is the effect of encouragement? What is the effect of doing what you are encouraged to do? The first question is comparatively easy to answer in a randomized encouragement experiment because encouragement was randomized. So far as the first question goes, there is nothing special about encouragement as a treatment, and the exclusion restriction is not needed to answer the first question. The second question is more complex because the investigator cannot randomly assign an individual to quit smoking. Quitting smoking is not easy, and the person who succeeds may be quite different from the person who fails. The exclusion restriction is relevant to the second question. For the second question, the investigator has randomized something, but not the right thing. Is that useful?

For instance, Judson Brewer and colleagues conducted a randomized trial that compared two programs designed to encourage and enable smoking cessation: a new program emphasizing mindfulness training (MT) and the American Lung Association's standard "freedom from smoking" (FFS) program. They concluded, "Compared to those randomized to the FFS intervention, individuals who received MT showed a greater rate of reduction in cigarette use during

What is the effect of encouragement? What is the effect of doing what you are encouraged to do? For the second question, the investigator has randomized something, but not the right thing. Is that useful?

treatment and maintained these gains during follow-up."[15] The researchers were answering the first question: What is the difference between MT and FFS, the difference between two forms of encouragement to quit smoking? They concluded that one form of encouragement was more effective than the other. Their study is simply a randomized trial in which the treatment happens to be different ways of encouraging a change in behavior. For the first question—on the effect of encouragement—there is nothing special about a randomized encouragement experiment.

What would happen if you tried to answer the second question? If you were the sort of person who responds to encouragement by quitting smoking, would quitting cause your lung function to improve? The experiment did not—and could not—randomize quitting, but it is a perfectly reasonable question.

What would you expect to see in the encouragement experiment that is relevant to the second question? If quitting improved lung function, and if more people quit under treatment MT than under treatment FFS, then you expect to see better lung function in the MT group. On the other hand, in both the MT and FFS groups, only a fraction of people in the experiment were abstinent seventeen weeks after the start of treatment, specifically 31 percent in the MT group and 6 percent in the FFS group. It is hard to quit, and most people did not. Presumably, the effect caused by quitting is much larger than the effect caused by better encouragement to quit because many people do not quit even with better encouragement. Does this line of reasoning make any sense? Or more precisely, when does it make sense and when does it not? What assumptions are hidden within this line of reasoning? If those assumptions were made explicit, would they yield more than an intuition about the effect of quitting being bigger than the effect of encouragement to quit? If the assumptions were explicit, would they yield an estimate of the effect caused by quitting? Is it absurd to say that MT caused 31 percent − 6 percent = 25 percent more people to quit smoking, or one in four more people to quit, so the effect of quitting on lung function should be four times larger than that of receiving MT rather than FFS? Is it absurd to say that the effect of encouragement is a diluted effect of actually quitting because many people do not quit, but if we allowed for dilution, then we would see the effect of quitting? It is not absurd, but a bit more is needed for this line of reasoning to make sense.

Instrumental Variables and the Complier Average Causal Effect

Consider a simplified version of encouragement to quit smoking in which people are either encouraged to quit or not, and they either quit or do not alter their smoking behavior. This is simpler than the actual study by Brewer and colleagues, in part because it excludes the possibility of smoking less without quitting. With this minor simplification and a few others, the current section will describe an important result about instrumental variables and the so-called complier average causal effect. The result is due to Joshua Angrist, Guido Imbens, and Donald Rubin.[16]

Return to Kim and James in chapter 1, where Kim has lung function r_{Tk} if encouraged or lung function r_{Ck} if not encouraged, while James has lung function r_{Tj} if encouraged and lung function r_{Cj} if not encouraged. The effect of encouragement on Kim is $r_{Tk}-r_{Ck}$ and the effect on James is $r_{Tj}-r_{Cj}$, so for the two of them together, the average effect of encouragement on lung function is ATE = $(1/2)$ $(r_{Tk}-r_{Ck} + r_{Tj}-r_{Cj})$. If there were many people like Kim and James, then in a similar way the average effect of encouragement, the ATE, is the average over all of them. Suppose we picked half of those many people at random, assigning them to encouragement and withholding encouragement from the rest. In chapters 1–2, we saw that the difference in the mean lung function in the encouraged and unencouraged groups is a good estimate of the average effect of encouragement, the ATE, on lung function in a large randomized experiment. This is all just as it was in chapters 1–2 because the ATE is the average effect of encouragement, and encouragement is a treatment that was randomized. There is nothing new, so far.

In addition to lung function, either Kim or James may quit smoking, with or without encouragement. Quitting smoking is just another outcome, and the same considerations apply to this second outcome. Quitting is represented by a 1, not quitting by a 0. Then Kim has $q_{Tk} = 1$ if she would quit if encouraged, $q_{Tk} = 0$ if she would not quit if encouraged, $q_{Ck} = 1$ if she would quit if not encouraged, and $q_{Ck} = 0$ if she would not quit if not encouraged. The same is true for James with his q_{Tj} if encouraged or q_{Cj} if not encouraged. For Kim and James, the average effect of encouragement on quitting is ATEq = $(1/2) (q_{Tk} - q_{Ck} + q_{Tj} - q_{Cj})$, where the "q" has been appended to emphasize that ATEq is the effect of encouragement on quitting, as opposed to the ATE, which is the average effect of encouragement on lung function. With many more people in a large randomized trial, we can estimate the average effect of encouragement on quitting, ATEq. There is nothing new here either; all is as it was in chapters 1–2, but for a second outcome, namely quitting smoking.

So let us presume that a large randomized trial has given us a good estimate of the average effect of encouragement on lung function, ATE, and also a good estimate of the average effect of encouragement on quitting, ATEq. Chapters 1–2 showed that a large randomized trial can produce such estimates in a straightforward way. We now want to estimate the effect of quitting on lung function. Here we face the problem, not confronted in chapters 1–2, that the investigator does not control whether a person heeds encouragement and quits smoking, so the investigator cannot randomize individuals to quit or not. Is there any sense in which we are better off having randomized encouragement? Or are we back to square one, studying a treatment that has not been randomized, as in any observational study? Is randomizing the wrong thing better than randomizing nothing?

Two fairly small assumptions are needed to keep the argument as simple as possible. It would be a bit perverse to continue smoking if I encourage you to quit, but to quit if I say, "Do whatever you want; I don't care whether you quit." In symbols, if James had $q_{Tj} = 0$ and $q_{Cj} = 1$, then he would always do the opposite of what I encourage him to do. People can be stubborn or perverse—it happens all the time—but just to keep the argument simple, let us assume that no one is perverse in this sense. Let us talk about a generic individual i, possibly Kim, possibly James, or possibly someone else, where the experiment has I individuals, $i = 1, 2, \ldots, I$, as in chapter 1. It is hard to quit smoking, and some people will not succeed, with or without encouragement; such an individual i has $q_{Ti} = 0$ and $q_{Ci} = 0$, so $q_{Ti} - q_{Ci} = 0 - 0 = 0$, and encouragement does not affect that individual's smoking behavior. Other people decide to do something, and go and do it—they do not need encouragement to quit—and they have $q_{Ti} = 1$ and $q_{Ci} = 1$, so $q_{Ti} - q_{Ci} = 1 - 1 = 0$, and encouragement does not affect their smoking behavior either. Finally, there is the individual who needs encouragement to quit, the so-called complier. The complier quits if encouraged, $q_{Ti} = 1$, but cannot quit unaided by encouragement, $q_{Ci} = 0$, so encouragement does affect whether a complier quits, $q_{Ti} - q_{Ci} = 1 - 0 = 1$. Concisely, by assuming that no one acts perversely, we have assumed that encouragement is never a barrier to quitting, that $q_{Ti} \geq q_{Ci}$. That is the first small assumption.

If the first small assumption is true, then ATEq is the proportion of compliers, or the total number of compliers divided by I. That is, ATEq is the average over all I individuals of a quantity, $q_{Ti} - q_{Ci}$, that is 1 for compliers and 0 for everyone else. Using the ideas in chapter 2, we can get a very good estimate of the proportion of compliers, ATEq, in a large randomized experiment; just take the proportion of quitters in the treated group minus the proportion of quitters in the control group.

The second small assumption is that some people do heed encouragement. Perhaps these people are few in number, but some people are compliers, quitting smoking only if they are encouraged to do so. It is hard to quit smoking, so perhaps there are few compliers, but some individuals i do have $q_{Ti} - q_{Ci} = 1 - 0 = 1$. Combining the first and second small assumptions, we conclude that the average effect of encouragement on quitting, ATEq, is a positive number. Of course, ATEq cannot be negative because by the first small assumption, $q_{Ti} - q_{Ci} \geq 0$ for every individual i, and the average of nonnegative numbers cannot be negative. Also, ATEq cannot be zero by the second small assumption, namely $q_{Ti} - q_{Ci} = 1$ for some individuals.

Although ATEq must be positive, it could be small, close to zero, if few people quit smoking because they are encouraged to do so. If everyone did what they are encouraged to do—if everyone was a complier—then ATEq = 100 percent, and by randomizing encouragement we are effectively randomized quitting. At seventeen weeks, Brewer and colleagues estimated ATEq to be 31 percent – 6 percent = 25 percent, or 25 percent compliers, as noted previously; that is, according to this estimate, 25 percent were caused to quit by virtue of receiving the more effective form of encouragement. If ATEq is not zero, the quantity ATE/ATEq does not involve dividing by zero. If we have good estimates of the numerator, ATE, and the denominator, ATEq, in a large randomized trial, then we can estimate the ratio, ATE/ATEq. If ATEq = 25 percent = 0.25, then ATE/ATEq = ATE/0.25 = 4 × ATE, which was the quantity mentioned in the previous section as perhaps the effect of quitting smoking on lung function. As you'll recall, in the previous section we wondered if just multiplying by four is absurd. It says that if one in four people quit when encouraged to quit, then the effect of quitting is four times as large as the effect of encouragement to quit. We are still wondering whether that is absurd.

It is at this moment that the exclusion restriction becomes important; it was the third condition in the previous section. The exclusion restriction says that encouragement can affect an individual's lung function only if encouragement causes that individual to quit smoking. The familiar saying "no pain, no gain" is the exclusion restriction in one-syllable words. Talk is talk, and quitting is something else. Talk is talk; if you want something to happen, then when the talking is done, you have to do something. Talk might help you quit, and quitting might improve lung function, but talk without quitting does nothing for lung function—that is the meaning of the exclusion restriction in this context. In symbols, $q_{Ti} - q_{Ci} = 0$ implies $r_{Ti} - r_{Ci} = 0$; that is, no pain, $q_{Ti} - q_{Ci} = 0$, implies no gain, $r_{Ti} - r_{Ci} = 0$.

If the exclusion restriction is true, something miraculous happens. It takes a moment to understand what happens, and another to understand why that something is miraculous, but both moments are well spent.

By definition in chapter 1, the average effect of encouragement on lung function, ATE, is the total over I individuals i of $r_{Ti} - r_{Ci}$ divided by I. If the exclusion restriction is true, then $r_{Ti} - r_{Ci} = 0$ for every individual i who is not a complier—that is, every individual with $q_{Ti} - q_{Ci} = 0$. So the ATE is the total over compliers of $r_{Ti} - r_{Ci}$ divided by I. The average effect of encouragement on quitting, ATEq, is the total over I individuals i of $q_{Ti} - q_{Ci}$ divided by I, so it is the number of compliers divided by I. The numerator and denominator of ATE/ATEq both divide by I, which cancels, so we can stop talking about dividing by I. Then ATE/ATEq is the total of $r_{Ti} - r_{Ci}$ for compliers divided by the number of compliers, so ATE/ATEq is the average of $r_{Ti} - r_{Ci}$ for compliers. Compliers are people who quit smoking when encouraged to do so and thus for compliers, ATE/ATEq is the average effect on lung function of quitting smoking; it is the so-called complier average causal effect. And *that* is what we wanted all along: we wanted the effect of quitting, even though quitting was not randomized; only encouragement to quit was randomized. So what we wanted has materialized.

What is miraculous about all of this is that we cannot recognize a complier when we see one. If Kim is encouraged to quit and she does, then she might be a complier, but she might have quit anyway. In symbols, if Kim is encouraged to quit and she does, then we know $q_{Tk} = 1$, but we do not know q_{Ck}, so we do not know whether she is a complier with $q_{Tk} = 1$ and $q_{Ck} = 0$. In the same way, if James is not encouraged and does not quit, then he might be a complier, but perhaps he would not have quit even if he had been encouraged. If James is not encouraged and does not quit, then we know $q_{Cj} = 0$, but we do not know q_{Tj}, so we do not know whether James is a complier with $q_{Tj} = 1$ and $q_{Cj} = 0$. In light of this, it is quite remarkable that we can estimate the average effect of quitting on lung function for compliers, despite our total inability to recognize a complier when we see one.

This claim that ATE/ATEq is the average effect of quitting for compliers is sufficiently important that I will say it again in two other ways. First, the exclusion restriction—no pain, no gain—says that the entire effect of quitting on lung function in the ATE for all I individuals is coming from those individuals who are compliers. So ATE/ATEq attributes the entire effect on lung function to the compliers—that is, to the increase in quitters caused by encouragement, ATEq. If encouragement causes one in four people to quit, then the entire benefit for all I people is coming from these one in four people, so the average effect on these one in four people must be four times larger than the average effect on everyone. The dilution argument from the previous section is not absurd, but the argument does depend on the exclusion restriction.

Here is a second way to say it. We do not know who is a complier. The coin we flip to assign encouragement also does not know who is a complier; it comes up heads half the time for everyone, for compliers and everyone else. By definition, for a complier, randomizing encouragement is randomizing quitting; compliers do what they are encouraged to do. Hidden inside a big experiment that randomized encouragement is a smaller one that randomized quitting for compliers. Buried inside an experiment that randomized the wrong thing is a smaller one that randomized the right thing. We spent a lot of time randomizing encouragement to people who were ignoring us, people whose outcomes were unaffected by the encouragement that they ignored, but the real work occurred elsewhere. For compliers, when we randomized encouragement, we randomized quitting.

The Effect of Being Offered a Housing Voucher or the Effect of Accepting It

Recall from earlier in this chapter the natural experiment by Jacob and Ludwig in which applicants were offered housing vouchers based on their position on a randomized waiting list. The randomization of offers led to straight-forward estimates of the effect on employment and earnings of being offered a housing voucher. As is turns out, however, many applicants who received an offer turned it down. Perhaps a person received an offer of a housing subsidy and went out to look for attractive, affordable private housing, but discovered they could not afford what they wanted even with the aid of a subsidy.

The disincentive effects that Jacob and Ludwig were studying seem unlikely to materialize if the subsidy is offered but declined. After all, the disincentive effect is thought to operate when earning more means your housing subsidy is reduced, yet if you decline the subsidy, there is no subsidy to reduce. This situation is parallel to randomizing encouragement. The offer of a subsidy was randomized, but actually receiving a subsidy entails accepting it when it is offered, and that was not randomized. Quite possibly, people who decline an offered subsidy are different from those who accept it.

So Jacob and Ludwig estimated the complier average causal effect—that is, the effect of a housing subsidy on people who accept the subsidy only if it is offered by their randomized placement on the waiting list. The estimated decline in earnings and employment caused by accepting the subsidy is still small, but not trivially small, and it is between two and three times larger than the effect of the offer of a subsidy. Their paper concludes with an extended discussion of ways to design assistance programs without unintended disincentives.

Choosing Situations in Which Biases in Treatment Assignment Are Smaller

A natural experiment is an attempt to avoid bias in treatment assignment by finding some natural setting in which treatments are nearly randomized. Lotteries and

A natural experiment is an attempt to avoid bias in treatment assignment by finding some natural setting in which treatments are nearly randomized.

discontinuities are two examples. Sometimes the treatment we care about is not randomized but some form of encouragement to accept that treatment is randomized. Under certain conditions, randomized encouragement permits estimation of treatment effects for those who take the treatment only if they are encouraged.

REPLICATION, RESOLUTION, AND EVIDENCE FACTORS

There can be no instant—let alone mechanical—rationality.

—Imre Lakatos, "History of Science and Its Rational Reconstructions"

Replication Is Not Repetition

For ethical or practical reasons, randomization is infeasible in a particular context and the effects caused by a treatment are examined in an observational study. The outcome exhibited is associated with the treatment received, and that association persists after adjustment for measured covariates, as discussed in chapter 4. As in chapter 5, a sensitivity analysis shows that a minor unmeasured covariate could not begin to explain this association, yet what can guarantee that an unmeasured covariate is minor? An effort was made to locate circumstances in which these unmeasured biases are small, yielding a natural experiment of the type in chapter 7, yet what can guarantee that this effort was successful? The most plausible counterclaims were anticipated, and quasi-experimental devices invalidated these most plausible grounds for doubt, as in chapter 6, but unanticipated counterclaims continue to be raised. How is the matter resolved? How is interminable debate terminated?

Would it help to repeat the same study with new data? It might help if the principal source of uncertainty stemmed from a small sample size. If the initial study was large enough, however, and if the original investigators were competent and honest, then doing the same study again is likely to raise the same doubts again. Big data is the solution only if small data was the problem.

Repetition without Resolution

Consider an example of an unsuccessful attempt to resolve matters by replication. Does clinical treatment for addiction reduce the use of illegal narcotics such as heroin and cocaine? Between 1969 and 2000, evaluations based on three large data collection efforts claimed that treatment does reduce addiction. Each evaluation claimed to have further strengthened the evidence of effectiveness by replicating the previous evaluations. Each data set— the Drug Abuse Reporting Program (DARP), Treatment Outcomes Prospective Study (TOPS), and Drug Abuse Treatment Outcome Studies (DATOS)—had collected data on more than ten thousand people who had entered treatment. The competence and honesty of the investigators were not a matter of contention. And yet in an evaluation of these studies, the US National Academy of Sciences noted,

> The RAND study compares . . . drug use of members of the TOPS sample who completed their treatment programs with . . . drug use of TOPS subjects who began treatment but dropped out within 3 months. . . . Suppose, however, that treatment dropouts are more predisposed to drug use than are those who complete treatment. If dropouts are more severely addicted or less motivated or have fewer social supports than those who complete treatment, the observed differences in . . . drug use of dropouts and completers may reflect differences in characteristics of these two groups, not the effect of treatment programs. . . . The people who complete their treatment program may be those who are more likely to reduce their drug use, whether or not they receive treatment.[1]

The comparison of people who stay in treatment and those who drop out is not irrelevant; it is not by itself convincing, though, for the reasons the academy gave. Seeing the same pattern in three studies is little more convincing than seeing it once.

For a sequence of studies to be more convincing than any one of those studies, later ones must eliminate, reduce, or at least vary some of the biases that led to uncertainty about the earlier studies. A later study should be immune to some counterclaim to which an earlier study was vulnerable, even if the later study remains vulnerable to counterclaims. Persistence may be the most underrated of human virtues, but in science there is more to persistence than repeating yourself.

Varied Views of a Single Object

Contrast the studies of addiction in the previous section with the studies that terminated debate about smoking as a cause of lung cancer.

• Early studies showed that heavy smokers have much higher rates of lung cancer than people who never smoked.[2]

• Laboratory studies showed that substances in tobacco smoke cause cancer in laboratory mice.[3]

- When heavy smokers died of causes other than lung cancer, autopsies revealed they had precancerous lesions in their lungs infrequently found among autopsied nonsmokers.[4]

- Women increased their smoking in the wake of aggressive advertising suggesting that independent, liberated, slender, glamorous women would, of course, smoke. "You've come a long way baby," said the ad for Virginia Slims cigarettes. After an appropriate wait—a few decades—lung cancer rates among women increased dramatically, while rates for men did not increase.[5]

These studies are not immune to counterclaims. True, heavy smokers chose to smoke and may differ from nonsmokers. True, mice are not people. True, precancerous lesions in the deceased will never become cancer. True, women increased their smoking, but they also took jobs that had previously been dominated by men, and some of those jobs presented occupational risks to lungs.

Nonetheless, the studies of smoking and lung cancer are unlike DARP, TOPS, and DATOS. One explanation could explain DARP, TOPS, and DATOS as something other than an effect of clinical treatment for addiction. Perhaps that alternative explanation is incorrect, but one explanation could explain everything. For smoking and lung cancer, many unrelated explanations must conspire to give the false impression that smoking causes lung cancer.

Replication is not repetition. Successful replication of an observational study removes or varies some potential bias that forms a reasonable ground for doubting earlier studies.

Replication is not repetition. Successful replication of an observational study removes or varies some potential bias that forms a reasonable ground for doubting earlier studies.

Evidence Factors

Replication concerns a new comparison that resists some reasonable doubt raised about previous comparisons. If replication is not about obtaining more data but instead about a new and independent comparison, then ask, Can an observational study replicate itself? Can one study do several comparisons such that doubts about one comparison do not apply to another comparison? The topic has technical aspects that I will leave to the references, so let me just mention my favorite example.[6]

Are children affected by toxins at a workplace they never visit? Do parents exposed to lead at work also expose their children? Do parents bring lead home in their clothes and hair, thereby exposing their children? David Morton and colleagues asked this question about the children of employees of a factory in the US state of Oklahoma where lead was used in the manufacture of batteries.[7] All the parents were fathers. Morton and colleagues measured the level of lead in the blood of these children, comparing each child to a neighboring control child close in age.

There are three comparisons in figure 11. The "worker versus control" plot, on the left, compares the level of lead in the control children and the children of fathers exposed to lead in the battery factory. Children whose fathers worked in the battery factory had more lead in their blood than did the control children. That is comparison one.

Fathers who worked at the battery factory performed different jobs, some with greater exposure to lead than others. Does the father's degree of exposure to lead predict the lead in his child's blood? Morton and colleagues divided the children of the exposed fathers into three groups based on their father's level of exposure to lead at work: a high-exposure group (H), medium-exposure group (M), and low-exposure group (L). The middle panel of figure 11 shows the lead in a child's blood by the classification of the father's exposure to lead at work. Dads with higher exposure to lead had children with more lead in their blood.

The high-exposure group was further divided based on the father's hygiene on leaving the factory at the end of the workday. Hygiene was OK if the father showered and changed clothes, or at least changed clothes, before leaving work, and was poor if he did none of these things. The right panel of figure 11 shows the lead in the blood of the children of those fathers with high exposure, classified by hygiene. Poor hygiene of the father was associated with higher levels of lead in a child's blood.

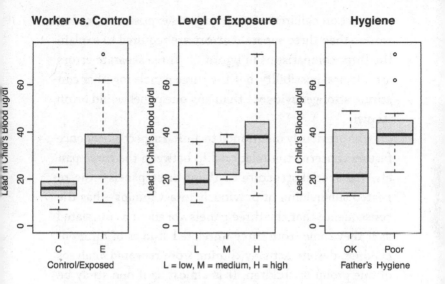

Figure 11 The left plot shows lead in the blood of control children (C) and children of fathers who were exposed to lead at work (E). The middle plot separates children in the E group by the father's level of exposure at work: low (L), medium (M), or high (H). The right plot separates children in the H group based on the father's hygiene when leaving work, as either OK or poor.

The simplest explanation of the patterns in figure 11 is that lead from the battery factory was affecting children who never set foot in the factory, being transported home by the father. Each of the three comparisons in the three panels of the figure is fallible in the way observational studies are always fallible. If the figure, however, does not mean what it appears to mean—if we are not looking at an effect on children of their father's exposure to lead at work—then three separate errors are required to explain the three comparisons in figure 11. Three separate errors are a logical possibility, but the three panels together constitute stronger evidence than any one panel would be on its own.

The previously mentioned technical aspect of evidence factors concerns the relationship between the three panels. The special structure of these three panels—the repeated subdividing of previously intact groups—has the consequence that the three panels are about as unrelated as if they came from three unrelated studies of different children, despite actually coming from repeated analyses of one group of children. It is almost as if one study became three studies, each susceptible to different biases.

UNCERTAINTY AND COMPLEXITY IN CAUSAL INFERENCE

And yet each of these stories had a virtue: as narratives, they seemed plausible, more than everyday or historical reality, which is far more complex and less credible. The stories seemed to explain something that was otherwise hard to understand.

—Umberto Eco, *Serendipities*

Are Small Daily Doses of Alcohol Beneficial?

Uncertainty and complexity are uncomfortable when a definite course of action is needed. What if the wrong action is chosen? No one welcomes uncertainty or puzzlement when faced with the prospect of a cruel illness followed by a slow death. Practical considerations like these can be a distraction when what the situation actually demands is some quiet time in the laboratory.

We are equally uncomfortable with certainty and simplicity, when clear evidence and plain logic dictate an unambiguous course of action that we would just as soon omit. Clear evidence and plain logic leading to an unwelcome conclusion can be such a downer.

Our sentiments can infect our evidence.

To maintain what Dewey called the experimental habit of mind, certain natural habits of mind need to be resisted.

The path of inquiry is often blocked by intolerance of uncertainty and complexity. Uncertainty and complexity must be acknowledged before they can be addressed.

I close this book with a brief discussion of causal inference about a topic of current controversy, namely the health benefits or harms of small daily doses of alcohol. I pick this topic with some trepidation. The topic is controversial and appropriately so, for it is complex, and extensive evidence has left much in doubt. Moreover, the topic may not remain controversial; it may be sorted out, perhaps sooner, perhaps later. Nonetheless, a causal inference assembles a jigsaw puzzle, and it is appropriate to close with one puzzle that is not fully assembled.

Oncologists versus Cardiologists

In 2018, oncologists threw down the gauntlet before cardiologists. The challenge consisted of an article and position

The path of inquiry is often blocked by intolerance of uncertainty and complexity. Uncertainty and complexity must be acknowledged before they can be addressed.

paper by Noelle LoConte and colleagues in the leading clinical cancer journal, the *Journal of Clinical Oncology*, titled "Alcohol and Cancer: A Statement of the American Society of Clinical Oncology." It is generally accepted in 2022 that high levels of alcohol consumption cause large numbers of deaths from various cancers, auto accidents, liver diseases, and violence. Drink a bottle of vodka a day—or even a bottle of wine a day—and your health will likely suffer.

Even moderate alcohol consumption by pregnant women poses risks to a developing fetus, with the risks being present before the pregnancy is known. The US Centers for Disease Control writes, "There is no known safe amount of alcohol use during pregnancy or while trying to get pregnant. There is also no safe time during pregnancy to drink."[1]

The active question in 2022 is whether a low daily dose of alcohol—the much-touted single glass of red wine at dinner—is harmful or beneficial. Does a single glass of wine lengthen or shorten life? The possible benefit is a reduced risk of death from cardiovascular disease. In their 2018 article in the *Journal of Clinical Oncology*, LoConte and colleagues wrote,

> Conflicting data about the heart health of alcohol, especially red wine, is one additional barrier to addressing alcohol and cancer risk. . . . Larger studies and meta-analyses have failed to show an all-cause

mortality benefit for low-volume alcohol use compared with abstinence or intermittent use, which suggests the lack of a true benefit to daily alcohol use. . . . [T]he benefit of alcohol consumption on cardiovascular health likely has been overstated. . . . [T]he risk of cancer is increased even with low levels of alcohol consumption, so the net effect of alcohol is harmful. Thus, alcohol consumption should not be recommended to prevent cardiovascular disease or all-cause mortality.[2]

It is easy to understand this perspective. Carcinogens are typically identified by studying people with high exposures, heavy smokers, uranium miners exposed to radon gas, and chemical workers exposed to benzene. Having demonstrated that a substance is a major cause of cancer at high doses in humans, we take prudent steps to minimize exposure. No one advocates smoking one cigarette per day simply because it is difficult to produce overwhelming evidence that a single cigarette is harmful.

Cardiologists have always been cautious when discussing alcohol. They invariably observe that some people face the choice of abstaining from alcohol or drinking excessively, and given that choice, abstaining is clearly better. Sylvain Tesson does well in expressing this thought: "It is at the fifth glass of vodka that resisting the next one becomes difficult."[3]

Are there cardiovascular benefits from low daily doses of alcohol? The evidence in favor of such a benefit is far from inconsequential. Moderate alcohol consumption is associated with a reduced risk of death from cardiovascular disease, together with an increase in HDL cholesterol, which is thought to be of benefit for cardiovascular disease.[4] Both associations have been produced repeatedly, over a long period of time, by independent investigators, and the two associations fit together nicely. Unlike the Russians, the French stubbornly refuse to die young.[5]

Still, cardiologists are hesitant to recommend even moderate consumption of wine. Speaking for the American Heart Association, Ira Goldberg and colleagues argued,

> Moderate intake of alcoholic beverages (1 to 2 drinks per day) is associated with a reduced risk of [coronary heart disease] in populations. . . . Despite the biological plausibility and observational data in this regard, it should be kept in mind that these are insufficient to prove causality. There are numerous examples in the cardiovascular literature of studies that have documented consistent population and mechanistic data that have not held up in clinical trials, e.g., β-carotene, vitamin E, and hormone replacement therapy. It is impossible to adequately adjust for factors related to human behavior that

cannot be measured in observational designs. Although moderate use of wine and other alcohol-containing beverages does not appear to lead to significant morbidity, alcohol ingestion, unlike other dietary modifications, poses a number of health hazards. Without a large-scale, randomized, clinical end-point trial of wine intake, there is little current justification to recommend alcohol (or wine specifically) as a cardioprotective strategy.[6]

A Dissenting Voice from a New Tactic: Mendelian Randomization

Not all studies find a reduction in cardiovascular risk from alcohol. A recent study by Michael Holmes and colleagues uses a newer tactic that suggests alcohol increases cardiovascular risk. The newer tactic is known as Mendelian randomization.[7]

Suppose that a variant of a certain gene made you drink less alcohol, that people received this variant at random, and that this gene affected cardiovascular disease only indirectly by reducing alcohol consumption. Admittedly, that is a fair amount of supposing, and what is being supposed may not be true. Nonetheless, if these suppositions were true—a big if—then the gene would be an instrument manipulating alcohol consumption, as in chapter 7. Mendelian randomization is interesting not because it is infallible—nothing outside a randomized trial comes close to infallible causal inference—but rather because its flaws are different from those that arise when treated and control groups are compared after adjustment for covariates. We believe an object is real, not an illusion, when we look at it from various fallible perspectives and keep seeing the same thing.

Holmes and colleagues noted,

> Carriers of the A-allele of ADH1B rs1229984 consumed 17.2% fewer units of alcohol per week . . . had a lower prevalence of binge drinking . . . and had higher abstention than non-carriers. . . . Individuals with a genetic variant associated with non-drinking and lower alcohol consumption had a more favorable cardiovascular profile and a reduced risk of coronary heart disease than those without the genetic variant. This suggests that reduction of alcohol consumption, even for light to moderate drinkers, is beneficial for cardiovascular health.[8]

Studies using Mendelian randomization face different problems than those that compare exposed and unexposed groups. As discussed in chapter 8, we hope to see that studies facing different problems concur about causal effects, and we saw that this did happen in the studies of smoking and lung cancer. In contrast, in this case, for light or moderate alcohol consumption, changing the study design, changing the methodology, changes the answer, from benefit to harm.

The Answer Might Be Complex

If there is, in fact, a trade-off of risk of cancer and risk of heart disease—a big if—then the trade-off might be complex. Due to different genes, people metabolize alcohol differently. Due to different genes, people face different risks of cancer. Due to different genes, people have different risks of heart disease. Humans have perhaps twenty-five thousand genes and consequential variants of many of them.

If there is a trade-off of risks of cancer and heart disease, then the best advice about a glass of red wine might be different for people with different genes. As I write in 2022, there is some understanding of how genes affect alcohol metabolism, cancer, and heart disease, but it is neither sufficiently precise nor complete to be the basis for subtle recommendations about a glass of red wine.

In part because alcohol seems to cause breast cancer, the US Centers for Disease Control does not advocate drinking but instead recommends "limiting intake to 2 drinks or less in a day for men or 1 drink or less in a day for women."[9] It is at least conceivable that other distinctions between people are also relevant besides sex.

A Traditional Toxin

Alcoholic beverages are older than written history. Homer's *Iliad* tells of intoxicated men fighting over a beautiful woman in the Middle East. Alcohol is present in ceremony and celebration. Friends, families, and lovers drink together. We drink to congratulate, commiserate, and drown our sorrows. There are friends you cannot have and places you cannot comfortably go if you abstain from alcohol. The world's religions use alcohol in their services or make it a mortal sin. Alcohol affects the brain, altering mood, judgment, competence, self-control, and behavior. A transient loss of self-control or competence can, and often does, have permanent, grave consequences.

Am I saying that our sentiments make it difficult to appraise evidence about the effects of alcohol? No. Alcohol is an example. I am saying that our sentiments make it difficult to appraise evidence.

A perfectly rational response to extant evidence is to look for ways to reduce cardiovascular risk without increasing cancer risk. Options are exercise, weight loss, changes in diet, more exercise. Rational though this may be, something else says that a treadmill is not quite adequate as a substitute for a glass of wine.

Total Mortality

Why is there conflict and debate about low daily doses of alcohol? In part, the debate is about total mortality as opposed to mortality from a specific disease. Many studies reporting benefits of low doses of alcohol focus on some aspect of mortality from coronary heart disease. If there is a decrease in the death rate from coronary heart disease—a big if—is it big enough to offset an increase in mortality from cancer, accidents, and other diseases? Would people live longer with or without a daily glass of wine? Is total mortality reduced? That is one question.

A study of total mortality is in closer contact with reality than is a study of mortality from specific diseases. People often tell model-based stories about reducing risk from one disease while holding other diseases constant. These stories have aspects that are not identified, aspects immune to data and evidence, like an argument over whether angels are vegetarians. This important fact about immunity from refutation was first demonstrated by Anastasios Tsiatis.[10]

Is Part or All of the Supposed Heart Benefit Simply a Mistake?

There is also concern that the ostensible benefits of alcohol for cardiovascular disease may be produced by a bias in who drinks a little, who drinks a lot, and who abstains. This possibility was raised in the 2018 article in the *Journal of Clinical Oncology* and is extensively discussed elsewhere.

This debate is about the so-called J-shaped curve, depicted in stylized form in figure 12.[11] The left panel (*a*) of figure 12 depicts a steady increase in mortality with an increasing daily alcohol consumption. The right panel (*b*) depicts the J-shaped curve, in which light alcohol consumption confers the lowest mortality. The debate is about which panel is a true depiction. Importantly, the "no benefit from alcohol" and "J-shaped curve" panels are not that different: there is high mortality at high levels of alcohol consumption in both panels, low mortality at low levels of consumption, and the difference between abstinence and light consumption is small by comparison in both panels. As we saw in chapter 5, small treatment effects are often sensitive to small biases. Both panels show dramatic harm from high levels of alcohol. An effect like that can often be distinguished from no effect by a series of well-executed observational studies; such an effect may be insensitive to small and moderate biases. In contrast, what distinguishes the panels are two different small effects at low

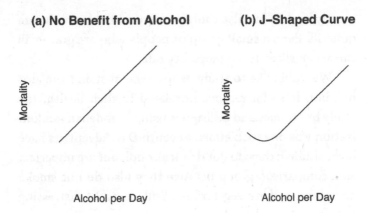

(a) No Benefit from Alcohol   **(b) J–Shaped Curve**

Mortality

Alcohol per Day

Mortality

Alcohol per Day

Figure 12 Stylized comparison of increasing mortality with increasing alcohol (*a*), and the so-called J-shaped curve in which the lowest mortality occurs with low alcohol consumption (*b*).

doses. Evidence that favors one panel of figure 12 over the other panel may be sensitive to small unmeasured biases.

What biases may exist? Well, why do some people abstain from alcohol, while others consume small daily doses? Some people, perhaps most people, abstain or not as a matter of taste. Others abstain for religious reasons. Some people abstain because they are recovering alcoholics and to drink even a little is to put their recovery in jeopardy. Some people abstain because of other illnesses that make alcohol consumption unwise. Some people abstain because they consume medications that preclude consuming alcohol. The small bend up in the right-hand panel of figure 12 may reflect a population of abstainers that is mostly healthy but also contains some individuals who are quite ill. Even a small group of people who are gravely ill can seriously distort a mortality rate.

We would like to study people who abstain from alcohol, but do so for reasons unrelated to their health. The study by Holmes and colleagues using Mendelian randomization was one such effort. Seventh Day Adventists have been studied; they do not drink alcohol, but are imperfect as a comparison group because they also do not smoke, and about half are vegetarians. Still, it can be distressing to learn how healthy they are.[12]

Of investigations of the possibility of bias, my favorite is a miniature study from 1982. The investigators set out to study the effects of alcohol, but demonstrated a bias instead. Bo Petersson, Erik Trell, and Hans Kristenson used ten structured questions to ask Swedish men about their alcohol consumption. One question asked about abstention from alcohol. Nine questions asked about the circumstances of alcohol consumption, perhaps problematic consumption. The lowest mortality was in the group that consumed some alcohol, but not much. Was the highest risk of death among those who consumed the most alcohol? No, the highest risk of death was among the abstainers. Why were the abstainers at such high risk? Petersson and colleagues wrote: "Most of these men, however, had chronic disease as the reason for their abstention, or even a past history of alcoholism. Increased mortality in non-drinkers may create a false impression of a preventive effect of any versus no daily drinking."[13]

We should not neglect the possibility that some people abstain from alcohol because they are ill as opposed to being ill because they abstain from alcohol. We are negligent if we fail to explore the reasons people abstain—that is, if we fail to compare outcomes for people who abstain for different reasons (see chapter 6). If abstention seems to lower the mortality of people who abstain for religious reasons, but seems to increase the mortality of people who abstain because they are recovering alcoholics, then that is evidence against harm caused by abstention. If both types of abstainers experience higher mortality than people who drink a glass of wine each day, then the failure to find evidence of bias where it might reasonably be expected to appear would lend some support to a claim that a glass of wine is beneficial. In general, philosopher Ernest Sosa observed, "If knowledge is a matter of apt intellectual performance in pursuit of truth [then] negligence can deny us knowledge. . . . We are negligent when we should be open to verifying evidence but close our minds instead. When this happens, our success is lucky. And the luck involved is no less luck when it is good luck."[14]

Although the article by Petersson, Trell, and Kristenson was published in a major medical journal almost forty years ago, it has not been heavily cited. It says that low doses of alcohol appear beneficial, but appearances seem to deceive. It implies that empirical science is difficult, and mistakes—at times grave ones—are common. It cautions that an observational study about a small treatment effect—the distinction between the two panels in figure 12—can easily give the wrong answer through negligence, overconfidence, or even unavoidable error. Who would cite an article that says so many true but unpleasant things?

So Are Small Daily Doses of Alcohol Beneficial?

Does a daily glass of red wine prolong or shorten life? What is the answer?

Time will tell.

Or maybe not.

Chapter 1: The Effects Caused by Treatments. The effect caused by a treatment is a comparison of two potential outcomes for the same person: the outcome that would be observed under treatment and the outcome that would be observed under control. Causal inference is difficult because a person receives either treatment or control, not both, so these two potential outcomes for one person are never both observed.

Chapter 2: Randomized Experiments. The problem in chapter 1 has no solution for a single person. A randomized experiment divides a finite population of people into treated and control groups by repeatedly flipping a fair coin to assign people to groups. When ethical and feasible, a randomized experiment solves the problem of inference about the typical effect of a treatment on that finite population of people.

Chapter 3: Observational Studies: The Problem. Random assignment to treatment or control is not always ethical, nor always feasible. In the absence of random assignment—that is, in an observational study—treated and control groups may differ prior to treatment in both ways we can see and cannot see. In an observational study, we risk mistaken conclusions about cause and effect by comparing treated and control groups that are not comparable.

Chapter 4: Adjustments for Measured Covariates.   When treated and control groups differ before treatment in a way we can see—when they differ in terms of measured covariates such as age—this problem can often be corrected by adjustments. The simplest form of adjustment is matched sampling: a control group is selected from a larger population of controls so that it resembles the treated group in terms of measured covariates—say, groups with similar ages. One tool useful in matching for many covariates is the propensity score, namely the probability of treatment for given observed covariates.

Chapter 5: Sensitivity to Unmeasured Covariates.   We can never be sure that treated and control groups differ only in ways we can see. The groups may differ in terms of unmeasured covariates. Would small differences that we cannot see alter the conclusions? The answer is provided by a sensitivity analysis.

Chapter 6: Quasi-Experimental Devices in the Design of Observational Studies.   Sensitivity analysis addresses small pretreatment differences that we cannot see, but of course the differences might be large, not small. Perhaps large differences that we cannot see leave behind visible traces that we can see. An invisible mouse may pass unnoticed, where an invisible elephant leaves a path of wreckage. Quasi-experimental devices, such as multiple control groups, are tools for discovering large unmeasured pretreatment differences between treated and control groups when such differences are present.

Chapter 7: Natural Experiments, Discontinuities, and Instruments. Random assignment of treatments is sometimes unethical or infeasible, but bits of randomness enter every life, altering its course. Natural experiments, discontinuities, and instruments are attempts to cull bits of natural randomness for constructive use in an observational study.

Chapter 8: Replication, Resolution, and Evidence Factors. Replication is central to science, but it is easy to use the same procedures to repeat the same mistakes. A series of observational studies may reach firmer conclusions than any one study in that series, providing successive studies are liable to different mistakes, not to repeating the same one. Evidence factors seek such replication within a single study through carefully varied comparisons.

Chapter 9: Uncertainty and Complexity in Causal Inference. Outside randomized experiments, causal inference is challenging. The path of inquiry is often blocked by intolerance of uncertainty and complexity. Uncertainty and complexity must be acknowledged before they can be addressed.

Average treatment effect (ATE)
In a population, the average treatment effect is the average of the causal effects for the individuals in that population. It is sometimes called the average causal effect (ACE). A randomized experiment provides a good estimate of the ATE. See **causal effect** and chapter 1.

Boxplot
In essence, a boxplot depicts many numbers using three numbers: the median, which cuts the data in half; the upper quartile, which cuts in half the data above the median; and the lower quartile, which cuts in half the data below the median. These three numbers define the box in a boxplot. The boxplot also shows, as points, individuals who are extreme. So-called whiskers are lines that extend up and down from the box to the largest and smallest individuals who are not extreme. See chapter 3.

Causal effect
A comparison of two potential outcomes for an individual, namely the outcome that the individual would exhibit if treated and the outcome that the individual would exhibit if assigned to control. See chapter 1.

Complier average causal effect
A complier is an individual who would accept the treatment if and only if encouraged to do so by the push of an instrument. The complier average causal effect is the ATE in the subpopulation of compliers. See **instrument** and chapter 7.

Covariate
A covariate describes an individual prior to treatment, so it is unaffected by the treatment that an individual later receives. An individual has two potential outcomes, one under treatment and the other under control, yet has only one value of a covariate. See chapter 2.

Instrument (or instrumental variable)
An instrument is a random push to accept the treatment rather than the control, where the push affects outcomes only if it succeeds in altering the treatment received. See **complier average causal effect** and chapter 7.

Natural experiment

A natural experiment is an observational study that exploits some naturally occurring but truly random form of treatment assignment. Treatments assigned by lottery are a typical example. See chapter 7.

Observational study

A study of the effects caused by a treatment in which individuals are not randomly assigned to treatment or control. Compare with randomized experiments. See chapter 3.

Propensity score

In an observational study, the propensity score is the probability of treatment for an individual with specified values of observed covariates. See chapters 3 and 4.

Quasi-experimental device

In an observational study, a quasi-experimental device is a data element added to the study design to shed light on a potential source of bias in treatment assignment. These devices are often simple yet enlightening and may undermine certain grounds for doubting the study's conclusions. Typical examples are multiple control groups, unaffected outcomes, and counterparts. See chapter 6.

Randomized experiment

In a randomized experiment or randomized trial, a truly random device, such as a coin flip, is used to assign one individual to treatment and another to control. See chapter 2.

Sensitivity analysis

Outside randomized experiments, association does not imply causation: a sufficiently large bias in treatment assignment can explain any association. A sensitivity analysis answers a practical question as it relates to a specific observational study. How much bias in treatment assignment would need to be present to explain away the association actually seen in this observational study? See chapter 5.

Chapter 1

1. John Rhodehamel, *George Washington: The Wonder of the Age* (New Haven, CT: Yale University Press, 2017), 78, 299.

2. Frank M. Snowden, *Epidemics and Society: From the Black Death to the Present* (New Haven, CT: Yale University Press, 2019), 17–18.

3. John Waller, *The Discovery of the Germ: Twenty Years That Transformed the Way We Think about Disease* (New York: Columbia University Press, 2002), 75–81.

4. John Dewey, *Reconstruction in Philosophy* (New York: Dover, 2004), 7.

5. George Pólya, *How to Solve It: A New Aspect of the Mathematical Method* (Princeton, NJ: Princeton University Press, 1973), 136.

6. Just rearranging the order of addition gives $r_{T_+} - r_{C_+} = (1/2)(r_{Tk} + r_{Tj}) - (1/2)(r_{Ck} + r_{Cj}) = (1/2)(r_{Tk} - r_{Ck}) + (1/2)(r_{Tj} - r_{Cj}) =$ ATE. Rearranging this again gives ATE $= (1/2)(r_{Tk} - r_{Cj}) + (1/2)(r_{Tj} - r_{Ck})$, which is the average over heads and tails, where heads gives Kim the treatment and James the control, and tails gives James the treatment and Kim the control. Similar rearrangements work with more than two individuals.

Chapter 2

1. World Health Organization, "Ebola Virus Disease," February 23, 2021, https://www.who.int/news-room/fact-sheets/detail/ebola-virus-disease.

2. For these omitted features, see Sabue Mulangu, Lori E. Dodd, Richard T. Davey Jr., Olivier Tshiani Mbaya, Michael Proschan, Daniel Mukadi, Mariano Lusakibanza Manzo, et al., "A Randomized, Controlled Trial of Ebola Virus Disease Therapeutics," *New England Journal of Medicine* 381, no. 24 (2019): 2293–2303; Michael A. Proschan, Lori E. Dodd, and Dionne Price, "Statistical Considerations for a Trial of Ebola Virus Disease Therapeutics," *Clinical Trials* 13, no. 1 (2016): 39–48.

3. Alex John London, "Equipoise in Research: Integrating Ethics and Science in Human Research," *Journal of the American Medical Association* 317, no. 5 (2017): 525.

4. John P. Gilbert, Richard J. Light, and Frederick Mosteller, "Assessing Social Innovations: An Empirical Base for Policy," in *Evaluation and Experiment: Some Critical Issues in Assessing Social Programs*, ed. Carl A. Bennett and Arthur A. Lumsdaine (New York: Academic Press, 1975), 149–150.

5. In this paragraph and elsewhere, I gloss over a minor technical issue: flipping a fair coin 343 times is not quite the same as picking 179 patients at random from 343 patients. See William G. Cochran, *Sampling Techniques* (New York: John Wiley and Sons, 1977). As this is not a technical book, I gloss over some other minor technical issues later without additional footnotes stating this fact.

6. Cochran, *Sampling Techniques*, chapter 2.

Chapter 3

1. M. Tirmarche, A. Raphalen, F. Allin, J. Chameaud, and P. Bredon, "Mortality of a Cohort of French Uranium Miners Exposed to Relatively Low Radon Concentrations," *British Journal of Cancer* 67, no. 5 (1993): 1090–1097.

2. William G. Cochran, "The Planning of Observational Studies of Human Populations (with Discussion)," *Journal of the Royal Statistical Society* A 128, no. 2 (1965): 234.

3. David Card and Alan B. Krueger, "Minimum Wages and Employment: A Case Study of the Fast-food Industry in New Jersey and Pennsylvania," *American Economic Review* 84, no. 4 (1994): 772–793.

4. José R. Zubizarreta, Magdalena Cerdá, and Paul R. Rosenbaum, "Effect of the 2010 Chilean Earthquake on Posttraumatic Stress," *Epidemiology* 24, no. 1 (2013): 79–87.

5. Jeffrey Milyo and Joel Waldfogel, "The Effect of Price Advertising on Prices: Evidence in the Wake of 44 Liquormart," *American Economic Review* 89, no. 5 (1999): 1081–1096; James W. Marquart and Jonathan R. Sorensen, "Institutional and Postrelease Behavior of Furman-Commuted Inmates in Texas," *Criminology* 26, no. 4 (1988): 677–694.

6. For a discussion of this example in greater detail with additional references, see in Paul R. Rosenbaum, *Design of Observational Studies* (New York: Springer, 2020), chapters 19–20.

7. David R. Cox and E. Joyce Snell, *Analysis of Binary Data* (New York: Chapman and Hall/CRC, 1989).

8. Dimitra Kale, Kaidy Stautz, and Andrew Cooper, "Impulsivity Related Personality Traits and Cigarette Smoking in Adults: A Meta-analysis Using the UPPS-P Model of Impulsivity and Reward Sensitivity," *Drug and Alcohol Dependence* 185 (2018): 149–167; Jane E. Sarginson, Joel D. Killen, Laura C. Lazzeroni, Stephen P. Fortmann, Heather S. Ryan, Alan F. Schatzberg, and Greer M. Murphy Jr., "Markers in the 15q24 Nicotinic Receptor Subunit Gene Cluster (CHRNA5-A3-B4) Predict Severity of Nicotine Addiction and Response

to Smoking Cessation Therapy," *American Journal of Medical Genetics Part B: Neuropsychiatric Genetics* 156, no. 3 (2011): 275–284.

Chapter 4

1. Paul R. Rosenbaum and Donald B. Rubin, "The Central Role of the Propensity Score in Observational Studies for Causal Effects," *Biometrika* 70, no. 1 (1983): 41–55.

2. Paul R. Rosenbaum, *Design of Observational Studies* (New York: Springer, 2020), chapter 11.

Chapter 5

1. Avner Baz, *When Words Are Called For* (Cambridge, MA: Harvard University Press, 2012), chapter 4.

2. Allan M. Brandt, *The Cigarette Century: The Rise, Fall, and Deadly Persistence of the Product That Defined America* (New York: Basic Books, 2007), chapter 12.

3. Irwin D. J. Bross, "Statistical Criticism," *Cancer* 13, no. 2 (1960): 394.

4. Ludwig Wittgenstein, *On Certainty* (New York: Harper and Row, 1969), #122.

5. Bross, "Statistical Criticism," 399.

6. Richard Doll and A. Bradford Hill, "The Mortality of Doctors in Relation to Their Smoking Habits," *British Medical Journal* 1, no. 4877 (1954): 1451–1455; E. Cuyler Hammond and Daniel Horn, "Smoking and Death Rates: Report on Forty-Four Months of Follow-up of 187,783 Men. 2. Death Rates by Cause," *Journal of the American Medical Association* 166, no. 11 (1958): 1294–1308; E. Cuyler Hammond, "Smoking in Relation to Mortality and Morbidity: Findings in the First Thirty-Four Months of Follow-up in a Prospective Study Started in 1959," *Journal of the National Cancer Institute* 32, no. 5 (1964): 1161–1188.

7. W. L. Laurence, "Cigarette-Cancer Links Disputed," *New York Times*, December 29, 1957, 101.

8. Ralph Waldo Emerson, *Essays and Poems* (London: Everyman, 1995), 29.

9. Jerome Cornfield, William Haenszel, E. Cuyler Hammond, Abraham M. Lilienfeld, Michael B. Shimkin, and Ernst L. Wynder, "Smoking and Lung Cancer: Recent Evidence and a Discussion of Some Questions," *Journal of the National Cancer Institute* 22, no. 1 (1959): 193.

10. Joel B. Greenhouse, "Commentary: Cornfield, Epidemiology and Causality," *International Journal of Epidemiology* 38, no. 5 (2009): 1200.

11. Paul R. Rosenbaum, *Design of Observational Studies* (New York: Springer, 2020), chapter 3.

12. For smoking and lung cancer, see Paul R. Rosenbaum, *Observational Studies* (New York: Springer, 2002), 115, 129. For seat belts in car crashes, see Rosenbaum, *Design of Observational Studies*, 182.

13. Ruoqi Yu, Dylan S. Small, and Paul R. Rosenbaum, "The Information in Covariate Imbalance in Studies of Hormone Replacement Therapy," *Annals of Applied Statistics* 15, no. 4 (2021): 2023–2042.

14. Emerson, *Essays and Poems*, 121.

Chapter 6

1. Wayne A. Ray, Katherine T. Murray, Kathi Hall, Patrick G. Arbogast, and C. Michael Stein, "Azithromycin and the Risk of Cardiovascular Death," *New England Journal of Medicine* 366, no. 20 (2012): 1881–1890.

2. Donald T. Campbell, *Methodology and Epistemology for Social Science: Selected Papers 1957–1986* (Chicago: University of Chicago Press, 1988), 177–179.

Chapter 7

1. Will Dobbie and Roland G. Fryer Jr., "The Medium-Term Impacts of High-Achieving Charter Schools," *Journal of Political Economy* 123, no. 5 (2015): 985–1037; Atila Abdulkadiroğlu, Parag A. Pathak, and Christopher R. Walters, "Free to Choose: Can School Choice Reduce Student Achievement?," *American Economic Journal: Applied Economics* 10, no. 1 (2018): 175–206.

2. Brian A. Jacob and Jens Ludwig, "The Effects of Housing Assistance on Labor Supply: Evidence from a Voucher Lottery," *American Economic Review* 102, no. 1 (2012): 272–304.

3. Scott Hankins, Mark Hoekstra, and Paige Marta Skiba, "The Ticket to Easy Street? The Financial Consequences of Winning the Lottery," *Review of Economics and Statistics* 93, no. 3 (2011): 961–969.

4. Jacob and Ludwig, "The Effects of Housing Assistance on Labor Supply," 273.

5. Bijayeswar Vaidya, Helen Imrie, Petros Perros, Eric T. Young, William F. Kelly, David Carr, David M. Large, et al., "The Cytotoxic T Lymphocyte Antigen-4 Is a Major Graves' Disease Locus," *Human Molecular Genetics* 8, no. 7 (1999): 1195–1199.

6. David Curtis, "Use of Siblings as Controls in Case-Control Association Studies," *Annals of Human Genetics* 61, no. 4 (1997): 319–333; Richard S. Spielman and Warren J. Ewens, "A Sibship Test for Linkage in the Presence of Association: The Sib Transmission/Disequilibrium Test," *American Journal of Human Genetics* 62, no. 2 (1998): 450–458.

7. Stavra N. Romas, Vincent Santana, Jennifer Williamson, Alejandra Ciappa, Joseph H. Lee, Haydee Z. Rondon, Pedro Estevez, et al., "Familial Alzheimer

Disease among Caribbean Hispanics: A Reexamination of Its Association with APOE," *Archives of Neurology* 59, no. 1 (2002): 87–91.

8. Richard S. Spielman, Ralph E. McGinnis, and Warren J. Ewens, "Transmission Test for Linkage Disequilibrium: The Insulin Gene Region and Insulin-Dependent Diabetes Mellitus (IDDM)," *American Journal of Human Genetics* 52, no. 3 (1993): 506–516.

9. Joseph D. Dougherty, Susan E. Maloney, David F. Wozniak, Michael A. Rieger, Lisa Sonnenblick, Giovanni Coppola, Nathaniel G. Mahieu, et al., "The Disruption of Celf6, a Gene Identified by Translational Profiling of Serotonergic Neurons, Results in Autism-Related Behaviors," *Journal of Neuroscience* 33, no. 7 (2013): 2732–2753.

10. Donald L. Thistlethwaite and Donald T. Campbell, "Regression-Discontinuity Analysis: An Alternative to the Ex Post Facto Experiment," *Journal of Educational Psychology* 51, no. 6 (1960): 309–317.

11. John DiNardo and David S. Lee, "Economic Impacts of New Unionization on Private Sector Employers: 1984–2001," *Quarterly Journal of Economics* 119, no. 4 (2004): 1385.

12. Sandra E. Black, "Do Better Schools Matter? Parental Valuation of Elementary Education," *Quarterly Journal of Economics* 114, no. 2 (1999): 577–599.

13. Luke Keele, Rocio Titiunik, and José R. Zubizarreta, "Enhancing a Geographic Regression Discontinuity Design through Matching to Estimate the Effect of Ballot Initiatives on Voter Turnout," *Journal of the Royal Statistical Society, series* A (2015): 223–239.

14. Paul W. Holland, "Causal Inference, Path Analysis and Recursive Structural Equations Models," *Sociological Methodology* 18 (1988): 449–484.

15. Judson A. Brewer, Sarah Mallik, Theresa A. Babuscio, Charla Nich, Hayley E. Johnson, Cameron M. Deleone, Candace A. Minnix-Cotton, et al., "Mindfulness Training for Smoking Cessation: Results from a Randomized Controlled Trial," *Drug and Alcohol Dependence* 119, no. 1–2 (2011): 72.

16. Joshua D. Angrist, Guido W. Imbens, and Donald B. Rubin, "Identification of Causal Effects Using Instrumental Variables," *Journal of the American Statistical Association* 91, no. 434 (1996): 444–455.

Chapter 8

1. Charles F. Manski, John V. Pepper, Yonette F. Thomas, and the US National Research Council, *Assessment of Two Cost-Effectiveness Studies on Cocaine Control Policy* (Washington, DC: National Academies Press, 1999), 17–18.

2. Richard Doll and A. Bradford Hill, "The Mortality of Doctors in Relation to Their Smoking Habits," *British Medical Journal* 1, no. 4877 (1954): 1451–1455;

Cuyler E. Hammond and Daniel Horn, "Smoking and Death Rates: Report on Forty-Four Months of Follow-up of 187,783 Men, 2: Death Rates by Cause," *Journal of the American Medical Association* 166, no. 11 (1958): 1294–1308.

3. E. Boyland, F. J. C. Roe, and J. W. Gorrod, "Induction of Pulmonary Tumours in Mice by Nitrosonornicotine, a Possible Constituent of Tobacco Smoke," *Nature* 202, no. 4937 (1964): 1126.

4. Oscar E. Auerbach, Cuyler Hammond, and Lawrence Garfinkel, "Changes in Bronchial Epithelium in Relation to Cigarette Smoking, 1955–1960 vs. 1970–1977," *New England Journal of Medicine* 300, no. 8 (1979): 381–386.

5. John C. Bailar and Heather L. Gornik, "Cancer Undefeated," *New England Journal of Medicine* 336, no. 22 (1997): 1569–1574.

6. Paul R. Rosenbaum, *Replication and Evidence Factors in Observational Studies* (New York: Chapman and Hall/CRC, 2021).

7. David E. Morton, Alfred J. Saah, Stanley L. Silberg, Willis L. Owens, Mark A. Roberts, and Marylou D. Saah, "Lead Absorption in Children of Employees in a Lead-Related Industry," *American Journal of Epidemiology* 115, no. 4 (1982): 549–555.

Chapter 9

1. "Alcohol Use in Pregnancy," Centers for Disease Control and Prevention, December 14, 2021, https://www.cdc.gov/ncbddd/fasd/alcohol-use.html.

2. Noelle K. LoConte, Abenaa M. Brewster, Judith S. Kaur, Janette K. Merrill, and Anthony J. Alberg, "Alcohol and Cancer: A Statement of the American Society of Clinical Oncology," *Journal of Clinical Oncology* 36, no. 1 (2018): 88.

3. Sylvain Tesson, *The Consolations of the Forest* (New York: Rizzoli, 2013), 140.

4. I. L. Suh, B. Jessica Shaten, Jeffrey A. Cutler, and Lewis H. Kuller, "Alcohol Use and Mortality from Coronary Heart Disease: The Role of High-Density Lipoprotein Cholesterol," *Annals of Internal Medicine* 116, no. 11 (1992): 881–887; Simona Costanzo, Augusto Di Castelnuovo, Maria Benedetta Donati, Licia Iacoviello, and Giovanni de Gaetano, "Alcohol Consumption and Mortality in Patients with Cardiovascular Disease: A Meta-analysis," *Journal of the American College of Cardiology* 55, no. 13 (2010): 1339–1347.

5. A. S. St. Leger, A. L. Cochrane, and F. Moore, "Factors Associated with Cardiac Mortality in Developed Countries with Particular Reference to the Consumption of Wine," *Lancet* 313, no. 8124 (1979): 1017–1020; S. de Renaud and Michel de Lorgeril, "Wine, Alcohol, Platelets, and the French Paradox for Coronary Heart Disease," *Lancet* 339, no. 8808 (1992): 1523–1526.

6. Ira J. Goldberg, Lori Mosca, Mariann R. Piano, and Edward A. Fisher, "Wine and Your Heart: A Science Advisory for Healthcare Professionals from the

Nutrition Committee, Council on Epidemiology and Prevention, and Council on Cardiovascular Nursing of the American Heart Association," *Circulation* 103, no. 3 (2001): 474.

7. George Davey Smith, and Shah Ebrahim, "Mendelian Randomization: Can Genetic Epidemiology Contribute to Understanding Environmental Determinants of Disease?," *International Journal of Epidemiology* 32, no. 1 (2003): 1–22. Tyler J. VanderWeele, Eric J. Tchetgen Tchetgen, Marilyn Cornelis, and Peter Kraft, "Methodological Challenges in Mendelian Randomization," *Epidemiology* 25, no. 3 (2014): 427–435.

8. Michael V. Holmes, Caroline E. Dale, Luisa Zuccolo, Richard J. Silverwood, Yiran Guo, Zheng Ye, David Prieto-Merino, et al., "Association between Alcohol and Cardiovascular Disease: Mendelian Randomisation Analysis Based on Individual Participant Data," *British Medical Journal* 349 (2014): g4164.

9. "Dietary Guidelines for Alcohol," Centers for Disease Control and Prevention, April 19, 2022, https://www.cdc.gov/alcohol/fact-sheets/moderate -drinking.htm. See also Chiara Scoccianti, Béatrice Lauby-Secretan, Pierre-Yves Bello, Véronique Chajes, and Isabelle Romieu, "Female Breast Cancer and Alcohol Consumption: A Review of the Literature," *American Journal of Preventive Medicine* 46, no. 3 (2014): S16–S25; Yin Cao, Walter C. Willett, Eric B. Rimm, Meir J. Stampfer, and Edward L. Giovannucci, "Light to Moderate Intake of Alcohol, Drinking Patterns, and Risk of Cancer: Results from Two Prospective US Cohort Studies," *British Medical Journal* 351 (2015): h4238.

10. Anastasios Tsiatis, "A Nonidentifiability Aspect of the Problem of Competing Risks." *Proceedings of the National Academy of Sciences* 72, no. 1 (1975): 20–22.

11. Michael Marmot and Eric Brunner, "Alcohol and Cardiovascular Disease: The Status of the U Shaped Curve," *British Medical Journal* 303, no. 6802 (1991): 565–568; Timothy Stockwell and Jinhui Zhao, "Alcohol's Contribution to Cancer Is Underestimated for Exactly the Same Reason That Its Contribution to Cardioprotection Is Overestimated," *Addiction* 112, no. 2 (2017): 230–232.

12. Roland L. Phillips, "Role of Life-style and Dietary Habits in Risk of Cancer among Seventh-Day Adventists," *Cancer Research* 35, no. 11, part 2 (1975): 3513–3522.

13. Bo Petersson, Erik Trell, and Hans Kristenson, "Alcohol Abstention and Premature Mortality in Middle-Aged Men," *British Medical Journal* 285, no. 6353 (1982): 1457–1459.

14. Ernest Sosa, *Epistemology* (Princeton, NJ: Princeton University Press, 2017), 169.

BIBLIOGRAPHY

Abdulkadiroğlu, Atila, Parag A. Pathak, and Christopher R. Walters. "Free to Choose: Can School Choice Reduce Student Achievement?" *American Economic Journal: Applied Economics* 10, no. 1 (2018): 175–206.

Angrist, Joshua D., Guido W. Imbens, and Donald B. Rubin. "Identification of Causal Effects Using Instrumental Variables." *Journal of the American Statistical Association* 91, no. 434 (1996): 444–455.

Auerbach, Oscar E., Cuyler Hammond, and Lawrence Garfinkel. "Changes in Bronchial Epithelium in Relation to Cigarette Smoking, 1955–1960 vs. 1970–1977." *New England Journal of Medicine* 300, no. 8 (1979): 381–386.

Bailar, John C., and Heather L. Gornik. "Cancer Undefeated." *New England Journal of Medicine* 336, no. 22 (1997): 1569–1574.

Black, Sandra E. "Do Better Schools Matter? Parental Valuation of Elementary Education." *Quarterly Journal of Economics* 114, no. 2 (1999): 577–599.

Boffetta, Paolo, and Mia Hashibe. "Alcohol and Cancer." *Lancet Oncology* 7, no. 2 (2006): 149–156.

Boyland, E., F. J. C. Roe, and J. W. Gorrod. "Induction of Pulmonary Tumours in Mice by Nitrosonornicotine, a Possible Constituent of Tobacco Smoke." *Nature* 202, no. 4937 (1964): 1126.

Brewer, Judson A., Sarah Mallik, Theresa A. Babuscio, Charla Nich, Hayley E. Johnson, Cameron M. Deleone, Candace A. Minnix-Cotton, et al. "Mindfulness Training for Smoking Cessation: Results from a Randomized Controlled Trial." *Drug and Alcohol Dependence* 119, no. 1–2 (2011): 72–80.

Bross, Irwin D. J. "Statistical Criticism." *Cancer* 13, no. 2 (1960): 394–400. Reprinted with eight commentaries in *Observational Studies* 4 (2018): 1–70.

Campbell, Donald T. *Methodology and Epistemology for Social Science: Selected Papers 1957–1986*. Chicago: University of Chicago Press, 1988.

Cao, Yin, Walter C. Willett, Eric B. Rimm, Meir J. Stampfer, and Edward L. Giovannucci. "Light to Moderate Intake of Alcohol, Drinking Patterns, and Risk of Cancer: Results from Two Prospective US Cohort Studies." *British Medical Journal* 351 (2015): h4238.

Card, David, and Alan B. Krueger. "Minimum Wages and Employment: A Case Study of the Fast-food Industry in New Jersey and Pennsylvania." *American Economic Review* 84, no. 4 (1994): 772–793.

Cochran, William G. "The Planning of Observational Studies of Human Populations (with Discussion)." *Journal of the Royal Statistical Society* A 128, no. 2 (1965): 234–266.

Cornfield, Jerome, William Haenszel, E. Cuyler Hammond, Abraham M. Lilienfeld, Michael B. Shimkin, and Ernst L. Wynder. "Smoking and Lung Cancer: Recent Evidence and a Discussion of Some Questions." *Journal of the National Cancer Institute* 22, no. 1 (1959): 173–203. Reprinted with commentaries by David R. Cox, Jan P. Vandenbroucke, Marcel Zwahlen and Joel B. Greenhouse in the *International Journal of Epidemiology* 38, no. 5 (2009): 1175–1191.

Costanzo, Simona, Augusto Di Castelnuovo, Maria Benedetta Donati, Licia Iacoviello, and Giovanni de Gaetano. "Alcohol Consumption and Mortality in Patients with Cardiovascular Disease: A Meta-analysis." *Journal of the American College of Cardiology* 55, no. 13 (2010): 1339–1347.

Cox, David R., and E. Joyce Snell. *Analysis of Binary Data*. New York: Chapman and Hall/CRC, 1989.

Curtis, David. "Use of Siblings as Controls in Case-Control Association Studies." *Annals of Human Genetics* 61, no. 4 (1997): 319–333.

Davey Smith, George, and Shah Ebrahim. "Mendelian Randomization: Can Genetic Epidemiology Contribute to Understanding Environmental Determinants of Disease?" *International Journal of Epidemiology* 32, no. 1 (2003): 1–22.

DiNardo, John, and David S. Lee. "Economic Impacts of New Unionization on Private Sector Employers: 1984–2001." *Quarterly Journal of Economics* 119, no. 4 (2004): 1383–1441.

Dobbie, Will, and Roland G. Fryer Jr. "The Medium-Term Impacts of High-Achieving Charter Schools." *Journal of Political Economy* 123, no. 5 (2015): 985–1037.

Doll, Richard, and A. Bradford Hill. "The Mortality of Doctors in Relation to Their Smoking Habits." *British Medical Journal* 1, no. 4877 (1954): 1451–1455.

Dougherty, Joseph D., Susan E. Maloney, David F. Wozniak, Michael A. Rieger, Lisa Sonnenblick, Giovanni Coppola, Nathaniel G. Mahieu, et al. "The

Disruption of Celf6, a Gene Identified by Translational Profiling of Serotonergic Neurons, Results in Autism-Related Behaviors." *Journal of Neuroscience* 33, no. 7 (2013): 2732–2753.

Eissa, Nada, and Jeffrey B. Liebman. "Labor Supply Response to the Earned Income Tax Credit." *Quarterly Journal of Economics* 111, no. 2 (1996): 605–637.

Fisher, Ronald A. *Design of Experiments*. Edinburgh: Oliver and Boyd, 1935.

Gastwirth, Joseph L. "Methods for Assessing the Sensitivity of Statistical Comparisons Used in Title VII Cases to Omitted Variables." *Jurimetrics Journal* 33 (1992): 19–34.

Gilbert, John P., Richard J. Light, and Frederick Mosteller. "Assessing Social Innovations: An Empirical Base for Policy." In *Evaluation and Experiment: Some Critical Issues in Assessing Social Programs*, edited by Carl A. Bennett and Arthur A. Lumsdaine, 39–193. New York: Academic Press, 1975.

Goldberg, Ira J., Lori Mosca, Mariann R. Piano, and Edward A. Fisher. "Wine and Your Heart: A Science Advisory for Healthcare Professionals from the Nutrition Committee, Council on Epidemiology and Prevention, and Council on Cardiovascular Nursing of the American Heart Association." *Circulation* 103, no. 3 (2001): 472–475.

Hammond, E. Cuyler. "Smoking in Relation to Mortality and Morbidity. Findings in the First Thirty-Four Months of Follow-up in a Prospective Study Started in 1959." *Journal of the National Cancer Institute* 32, no. 5 (1964): 1161–1188.

Hammond, E. Cuyler, and Daniel Horn. "Smoking and Death Rates: Report on Forty-Four Months of Follow-up of 187,783 Men. 2. Death Rates by Cause." *Journal of the American Medical Association* 166, no. 11 (1958): 1294–1308.

Hankins, Scott, Mark Hoekstra, and Paige Marta Skiba. "The Ticket to Easy Street? The Financial Consequences of Winning the Lottery." *Review of Economics and Statistics* 93, no. 3 (2011): 961–969.

Holland, Paul W. "Causal Inference, Path Analysis and Recursive Structural Equations Models." *Sociological Methodology* 18 (1988): 449–484.

Holmes, Michael V., Caroline E. Dale, Luisa Zuccolo, Richard J. Silverwood, Yiran Guo, Zheng Ye, David Prieto-Merino, et al. "Association between Alcohol and Cardiovascular Disease: Mendelian Randomisation Analysis Based on Individual Participant Data." *British Medical Journal* 349 (2014): g4164.

Jacob, Brian A., and Jens Ludwig. "The Effects of Housing Assistance on Labor Supply: Evidence from a Voucher Lottery." *American Economic Review* 102, no. 1 (2012): 272–304.

Keele, Luke, Rocio Titiunik, and José R. Zubizarreta. "Enhancing a Geographic Regression Discontinuity Design through Matching to Estimate the Effect of Ballot Initiatives on Voter Turnout." *Journal of the Royal Statistical Society*, series A (2015): 223–239.

Lawlor, Debbie A., Kate Tilling, and George Davey Smith. "Triangulation in Aetiological Epidemiology." *International Journal of Epidemiology* 45, no. 6 (2016): 1866–1886.

LoConte, Noelle K., Abenaa M. Brewster, Judith S. Kaur, Janette K. Merrill, and Anthony J. Alberg. "Alcohol and Cancer: A Statement of the American Society of Clinical Oncology." *Journal of Clinical Oncology* 36, no. 1 (2018): 83–93.

London, Alex John. "Equipoise in Research: Integrating Ethics and Science in Human Research." *Journal of the American Medical Association* 317, no. 5 (2017): 525–526.

Marmot, Michael, and Eric Brunner. "Alcohol and Cardiovascular Disease: The Status of the U Shaped Curve." *British Medical Journal* 303, no. 6802 (1991): 565–568.

Marquart, James W., and Jonathan R. Sorensen. "Institutional and Postrelease Behavior of Furman-Commuted Inmates in Texas." *Criminology* 26, no. 4 (1988): 677–694.

Milyo, Jeffrey, and Joel Waldfogel. "The Effect of Price Advertising on Prices: Evidence in the Wake of 44 Liquormart." *American Economic Review* 89, no. 5 (1999): 1081–1096.

Morton, David E., Alfred J. Saah, Stanley L. Silberg, Willis L. Owens, Mark A. Roberts, and Marylou D. Saah. "Lead Absorption in Children of Employees in a Lead-Related Industry." *American Journal of Epidemiology* 115, no. 4 (1982): 549–555.

Mulangu, Sabue, Lori E. Dodd, Richard T. Davey Jr., Olivier Tshiani Mbaya, Michael Proschan, Daniel Mukadi, Mariano Lusakibanza Manzo, et al. "A Randomized, Controlled Trial of Ebola Virus Disease Therapeutics." *New England Journal of Medicine* 381, no. 24 (2019): 2293–2303.

Neyman, Jerzy. "On the Application of Probability Theory to Agricultural Experiments. Essay on Principles." *Statistical Science* 5, no. 4 (1990): 465–480. English translation of an article published in Polish in 1923.

Petersson, Bo, Erik Trell, and Hans Kristenson. "Alcohol Abstention and Premature Mortality in Middle-Aged Men." *British Medical Journal* 285, no. 6353 (1982): 1457–1459.

Piantadosi, Steven. *Clinical Trials: A Methodologic Perspective*. New York: John Wiley and Sons, 2017.

Proschan, Michael A., Lori E. Dodd, and Dionne Price. "Statistical Considerations for a Trial of Ebola Virus Disease Therapeutics." *Clinical Trials* 13, no. 1 (2016): 39–48.

Ray, Wayne A., Katherine T. Murray, Kathi Hall, Patrick G. Arbogast, and C. Michael Stein. "Azithromycin and the Risk of Cardiovascular Death." *New England Journal of Medicine* 366, no. 20 (2012): 1881–1890.

Reichardt, Charles S. *Quasi-Experimentation*. New York: Guilford Publications, 2019.

Romas, Stavra N., Vincent Santana, Jennifer Williamson, Alejandra Ciappa, Joseph H. Lee, Haydee Z. Rondon, Pedro Estevez, et al. "Familial Alzheimer Disease among Caribbean Hispanics: A Reexamination of Its Association with APOE." *Archives of Neurology* 59, no. 1 (2002): 87–91.

Rosenbaum, Paul R. *Design of Observational Studies*. 2nd ed. New York: Springer, 2020.

Rosenbaum, Paul R. *Replication and Evidence Factors in Observational Studies*. New York: Chapman and Hall/CRC, 2021.

Rosenbaum, Paul R. "Sensitivity Analysis for Certain Permutation Inferences in Matched Observational Studies." *Biometrika* 74, no. 1 (1987): 13–26.

Rosenbaum, Paul R., and Donald B. Rubin. "The Central Role of the Propensity Score in Observational Studies for Causal Effects." *Biometrika* 70, no. 1 (1983): 41–55.

Rubin, Donald B. "Estimating Causal Effects of Treatments in Randomized and Nonrandomized Studies." *Journal of Educational Psychology* 66, no. 5 (1974): 688–701.

Scoccianti, Chiara, Béatrice Lauby-Secretan, Pierre-Yves Bello, Véronique Chajes, and Isabelle Romieu. "Female Breast Cancer and Alcohol Consumption: A Review of the Literature." *American Journal of Preventive Medicine* 46, no. 3 (2014): S16–S25.

Spielman, Richard S., and Warren J. Ewens. "A Sibship Test for Linkage in the Presence of Association: The Sib Transmission/Disequilibrium Test." *American Journal of Human Genetics* 62, no. 2 (1998): 450–458.

Spielman, Richard S., Ralph E. McGinnis, and Warren J. Ewens. "Transmission Test for Linkage Disequilibrium: The Insulin Gene Region and Insulin-Dependent Diabetes Mellitus (IDDM)." *American Journal of Human Genetics* 52, no. 3 (1993): 506–516.

St. Leger, A. S., A. L. Cochrane, and F. Moore. "Factors Associated with Cardiac Mortality in Developed Countries with Particular Reference to the Consumption of Wine." *Lancet* 313, no. 8124 (1979): 1017–1020.

Stockwell, Timothy, and Jinhui Zhao. "Alcohol's Contribution to Cancer Is Underestimated for Exactly the Same Reason That Its Contribution to Cardioprotection Is Overestimated." *Addiction* 112, no. 2 (2017): 230–232.

Stolley, Paul D. "When Genius Errs: RA Fisher and the Lung Cancer Controversy." *American Journal of Epidemiology* 133, no. 5 (1991): 416–425.

Suh, I. L., B. Jessica Shaten, Jeffrey A. Cutler, and Lewis H. Kuller. "Alcohol Use and Mortality from Coronary Heart Disease: The Role of High-Density Lipoprotein Cholesterol." *Annals of Internal Medicine* 116, no. 11 (1992): 881–887.

Thistlethwaite, Donald L., and Donald T. Campbell. "Regression-Discontinuity Analysis: An Alternative to the Ex Post Facto Experiment." *Journal of Educational Psychology* 51, no. 6 (1960): 309–317.

Tirmarche, M., A. Raphalen, F. Allin, J. Chameaud, and P. Bredon. "Mortality of a Cohort of French Uranium Miners Exposed to Relatively Low Radon Concentrations." *British Journal of Cancer* 67, no. 5 (1993): 1090–1097.

Tomar, Scott L., and Samira Asma. "Smoking-Attributable Periodontitis in the United States: Findings from NHANES III." *Journal of Periodontology* 71, no. 5 (2000): 743–751.

Tsiatis, Anastasios. "A Nonidentifiability Aspect of the Problem of Competing Risks." *Proceedings of the National Academy of Sciences* 72, no. 1 (1975): 20–22.

Tukey, John W. *Exploratory Data Analysis*. Waltham, MA: Addison-Wesley, 1977.

US Centers for Disease Control and Prevention. "Smoking, Gum Disease, and Tooth Loss." March 23, 2020. https://www.cdc.gov/tobacco/campaign/tips /diseases/periodontal-gum-disease.html.

US Surgeon General's Advisory Committee on Smoking. *Smoking and Health*. Washington, DC: US Department of Health, Education, and Welfare, Public Health Service, 1964.

Vaidya, Bijayeswar, Helen Imrie, Petros Perros, Eric T. Young, William F. Kelly, David Carr, David M. Large, et al. "The Cytotoxic T Lymphocyte Antigen-4 Is a Major Graves' Disease Locus." *Human Molecular Genetics* 8, no. 7 (1999): 1195–1199.

VanderWeele, Tyler J., Eric J. Tchetgen Tchetgen, Marilyn Cornelis, and Peter Kraft. "Methodological Challenges in Mendelian Randomization." *Epidemiology* 25, no. 3 (2014): 427–435.

Welch, B. L. "On the Z-Test in Randomized Blocks and Latin Squares." *Biometrika* 29, no. 1–2 (1937): 21–52.

Yu, Ruoqi, Dylan S. Small, and Paul R. Rosenbaum. "The Information in Covariate Imbalance in Studies of Hormone Replacement Therapy." *Annals of Applied Statistics* 15, no. 4 (2021): 2023–2042.

Zubizarreta, José R., Magdalena Cerdá, and Paul R. Rosenbaum. "Effect of the 2010 Chilean Earthquake on Posttraumatic Stress." *Epidemiology* 24, no. 1 (2013): 79–87.

FURTHER READING

Angrist, Joshua D., and Alan B. Krueger. "Empirical Strategies in Labor Economics." In *Handbook of Labor Economics*, edited by Orley Ashenfelter and David Card, 3:1277–1366. New York: Elsevier, 1999.

Cox, David R., and Nancy Reid. *The Theory of the Design of Experiments*. New York: Chapman and Hall/CRC, 2000.

Gerber, Alan S., and Donald P. Green. *Field Experiments: Design, Analysis, and Interpretation*. New York: W. W. Norton, 2012.

Hernán, Miguel A., and James M. Robins. *Causal Inference*. New York: Chapman and Hall/CRC, 2010.

Imbens, Guido W., and Donald B. Rubin. *Causal Inference in Statistics, Social, and Biomedical Sciences*. New York: Cambridge University Press, 2015.

Morgan, Stephen L., and Christopher Winship. *Counterfactuals and Causal Inference*. New York: Cambridge University Press, 2014.

Reichardt, Charles S. *Quasi-Experimentation*. New York: Guilford Publications, 2019.

Rosenbaum, Paul R. *Observation and Experiment: An Introduction to Causal Inference*. Cambridge, MA: Harvard University Press, 2017.

Rutter, Michael. *Identifying the Environmental Causes of Disease: How Should We Decide What to Believe and When to Take Action?* London: Academy of Medical Sciences, 2007. https://acmedsci.ac.uk/publications.

Shadish, William R., Thomas D. Cook, and Donald T. Campbell. *Experimental and Quasi-Experimental Designs for Generalized Causal Inference*. Boston: Houghton Mifflin, 2002.

PAUL R. ROSENBAUM is the Robert G. Putzel Professor Emeritus of Statistics and Data Science at the Wharton School of the University of Pennsylvania, where he worked from 1986 to 2021. For work in causal inference, he received the R. A. Fisher Award in 2019 and the George W. Snedecor Award in 2003 from the Committee of Presidents of Statistical Societies. He is the author of four other books, *Observational Studies*, *Design of Observational Studies*, *Observation and Experiment: An Introduction to Causal Inference*, and *Replication and Evidence Factors in Observational Studies*.